Research and Practice on the Theory of Inventive Problem Solving (TRIZ)

Leonid Chechurin
Editor

Research and Practice on the Theory of Inventive Problem Solving (TRIZ)

Linking Creativity, Engineering and Innovation

Springer

Editor
Leonid Chechurin
Lappeenranta University of Technology
Lappeenranta
Finland

ISBN 978-3-319-31780-9 ISBN 978-3-319-31782-3 (eBook)
DOI 10.1007/978-3-319-31782-3

Library of Congress Control Number: 2016947785

Printed on acid-free paper

This Springer imprint is published by Springer Nature
The registered company is Springer International Publishing AG Switzerland

Preface

We enjoy automation of more and more human activities. Automation enters the domain of analytical efforts: more and more elements of knowledge mining are turned into algorithms, for example, elements of modeling, optimization, information search and processing, etc. What has been an art becomes a standard routine, an algorithm realized in a software. But one fortress seems to stay bold and independent: it is still unclear how a new idea or new paradigm can be generated as the result of an algorithm. If it were possible, the conceptual design or invention could have been a controllable and predictable process. Computers could have generated new knowledge, new ideas, submit new research papers, and file new patents... Many efforts in artificial intelligence or literature-based discovery research are spent to mimic, to support, or to automate creative thinking, heuristic synthesis, and hypothesis generation.

The book contributes to the development and discussion on one of the most promising ideation tool: the theory for inventive problem solving (TRIZ). We invited an excellent crowd of TRIZ researchers and practitioners of different regions, backgrounds, and professions to share the thoughts and experience—to talk about possible evolution of the theory, its applications, and problems.

One more name can be found on the cover of the book; it is written with invisible ink. Prof. Alex Brem of The University of Southern Denmark has contributed much to this project. Prof. Brem suggested the idea of writing a book, set up the project with the publisher, invited some of the authors to contribute, and screened the contributions. At the same time, Prof Brem insisted on remaining outside the coeditor board, claiming that his contribution had been "not big enough." The editor expresses his great appreciation for his help and admires greatly his model example of scientific tenacity.

Lappeenranta, Finland — Leonid Chechurin
Spring 2016

Acknowledgments

The assistance of Iuliia Shnai, the MSc student of the Lappeenranta University of Technology, made the communication logistics between authors, reviewers, manuscripts, and editors much easier. Iuliia helped a lot with much of technical work.

The editor would also like to acknowledge the Finnish Innovation Agency TEKES and its FiDiPro program for its support.

Contents

Introduction

Leonid Chechurin

Abstract This editorial presents the motivation behind this book and gives an overview of the history of TRIZ, the academic research on the topic so far. The editorial perspectives in this chapter are based on almost 20 years' experience of activities in academia and industry where TRIZ was one of, but not the only, main subjects. The editor provides a special attention to TRIZ from the scientific perspective, elaborates on its weak and strong points, and discusses the current scientific landscape and perspectives. The chapter aims at assisting readers unfamiliar with TRIZ, to get acquainted of its history and context of application, structure, and advantages and to prepare for assimilating the chapters that follow, which could be challenging for beginners. Finally, the chapter briefly introduces all the contributions, linking the whole book in one.

Keywords TRIZ • Science • Overview

1 Motivation

Generally, it is a good idea to open the introduction by relevant definitions, which is in this case a definition of innovation. Innovation is a word that is applicable for almost anything new resulting from intentional efforts of a human. An "innovation tag" is suitable for a new product or new service; therefore, the word frequently decorates companies' profiles and advertisings, media breaking news titles, and business schools' education programs. Sometimes a process is called innovation, which is then a process of turning new knowledge into a new product (commercially successful if we talk about market-driven economy). Obviously, new knowledge or a new idea is a necessary part of innovation, but real innovation is more than that. An invention is to be given much more work before it is called innovation: marketing, management, financing, prototyping, manufacturing, and sale, among others. And for any new product, this process needs to be newly designed in order to be successful.

L. Chechurin (✉)
Lappeenranta University of Technology, LUT, Lappeenranta, Finland
e-mail: Leonid.Chechurin@lut.fi

L. Chechurin (ed.), *Research and Practice on the Theory of Inventive Problem Solving (TRIZ)*, DOI 10.1007/978-3-319-31782-3_1

Although most inventors don't mind to be called innovators, the biggest challenge of innovators doesn't seem to be finding the idea, but uncertainties and disturbances of the process of turning this idea into profit in the real world. If an analogy is allowed, the importance of new ideas for innovation is the same as the importance of bubbles for champaign.

Although nonmaterial as new ideas, bubbles are very important, even crucial components for champaign, but it is still just bubbles. Creativity is needed at all stages of the innovation process besides just new product conceptual design. Nonstandard schemes of investments can save the financing plan, creative market placement can increase product success, etc. But the stage of inventing a product is obviously the home court of creativity. Although innovation is a very popular word and a "must have term" to attract a bit more attention (consider the title of this book), to position TRIZ as *the* innovation tool is roughly the same as declaring a toothbrush as an instrument for body cleaning. A better fit would be calling TRIZ an instrument for inventing, ideation, idea generating. So, if it had not been for the popularity of the word innovation, a more precise title of this book would be about creativity and TRIZ in invention.

2 History

Genrich Altshuller introduced the elements of more productive thinking in inventive engineering in the USSR in 1956, in his paper coauthored by R. Shapiro (Altshuller and Shapiro 1956). Describing the ideation phase of engineering design more systematic and therefore gaining popularity among practicing inventors, the method evolved into a toolset for systematic creativity under the name "Theory of inventive problem solving" (TRIZ) in the 1980s and then "General theory of strong thinking" (OTSM) and "Lifetime strategy for creative persons" (ZhSTL) in the 1990s. G. Altshuller and his followers deployed TRIZ through extensive public activities, training seminars, articles, and books. TRIZ gained new instruments and chapters. The main method application roadmap, named the "Algorithm for inventive problem solving" (ARIZ), evolved through several editions from 1965 to 1985. The hype of education, inventing, engineering, and technological advance that existed in the USSR formed an excellent soil for the method to be of interest. Altshuller edited a column on creativity in the youth weekly newspaper *Pionerskaya Pravda* with a circulation of 9.5 million (nine and a half million!). I remember being a fan of the column as a kid.

Interestingly, that first publication of Altshuller in 1956 at the same time became his last publication in a scientific journal. He suffered a lot from the political regime in the USSR and therefore he decided that he would never work for governmental or state institutions, including schools and universities. And we should know that there were no other institutions in the USSR available until it collapsed. Writing science fiction books for living, Altshuller was never a member of a professional research community that used scientific publications as the primary stage for reporting results, discussion, development, and deployment of new knowledge. But he declared his findings as theory and the school he established with his followers pretended to research and to develop it further. Thus, unfortunately, the discussion, intentionally or not, never left the mostly closed circle of the TRIZ developers' community, and all the possible developments had to be approved by the founder rather than peer reviewed. In other words, the development of the "Theory for inventive problem solving" never entered the most traditional process for institutions and mechanisms of science.

According to one of his followers and colleagues, Vladimir Petrov, Altshuller was suggested to develop the findings into the form of a scientific dissertation, but he considered this framework as limiting and restricting. Obviously, Altshuller could not bear the conservatism of the academic society that developed new knowledge by small and cautious, but firm steps. He preferred a kind of shortcut, if a shortcut is possible on the way to distill new knowledge and to prove that a methodology works. The result can be seen as strongly nonlinear, which allowed quick development in the beginning, because no time was "wasted" on state-of-the-art analysis, careful experiment settings, peer reviewing, discussions, etc. But at the end of the day, it reduced the style and contents of the research to the level of publicism, school of thought, or conventional wisdom. We have to admit that a big share of deliverables of Altshuller and his followers were of speculative origin, based on or provided anecdotal evidence and could hardly be reproduced. These findings contain interesting, paradoxical, eye opening, and extremely useful insights for practice, but it is not enough research to be called science. In other words, more efforts are needed to develop TRIZ to a field of science and these efforts have been initiated relatively recently.

At the same time, many of these early developments have been proven to be useful in practice and therefore became a subject or instrument of current research activities (e.g., most "information technology + TRIZ" indexed papers or product design contributions use the function analysis approach. The latter appeared first in two patents (Tsourikov et al. 2000; Devoino et al. 2011)).

Is "theory" a legitimate word for TRIZ? Was Genrich Altshuller a scientist? Do his findings belong to science? These questions still provoke emotional discussions, taking into account that the definition of science is diverse. We can't help adding to these discussions and definitions one more paragraph.

We have to balance between these extremes. Science carefully delivers us new knowledge that becomes common good. This new knowledge might be correct but useless. We have to confess that sadly a big share of scientific research and publications is originated by points won by other publications ("publishing for publishing"). The practice is interested in knowledge that is applicable, whether this knowledge is well proven or not is of secondary interest. Thus, in some market-driven practices, such as consulting businesses, an ability to sell a theory proves its correctness. Even more, it shows that this is the best theory ever. Interestingly enough, business practice based on scientifically proven knowledge is the goal for most of the advanced universities nowadays. At the same time, reliable business is to be based on scientifically proven knowledge, for the sake of sustainability as well as reputation.

Genrich Altshuller enriched humankind with several insights of different values and application fields. For example, the trends for the engineering system evolution provide a systematic point of view on the past and future of products and technologies. From the same perspective, K. Marx enriched us by the systematic approach to observe the history of economic relations, J. Schumpeter by highlighting the innovation component in entrepreneurial competition, and D. Kondratieff by finding long-term periodicity in world economic index history record. All these examples are the insights of generic or philosophical depth. If the authors of these and similar approaches are called scientists and their theories are called science, the same applies to G. Altshuller and TRIZ.

At the same time, these influential insights remain a paradigm still, a school of thoughts rather than scientifically proven facts. Indeed, we have not yet come across any reliable proofs of Marxian capitalism nature or the evidence of long-term economic cycle existence (the original analysis of Kondratieff was based on 150 years of economic indicators' "Fourier transform" that yielded almost negligible long-term cycle of a period of 70 years; from the point of view of physics, the result is speculative; in other words it is too early to conclude that the long-term cycle exists; the analysis was repeated recently and still does not allow a sound conclusion). Thus, there has been no statistical research published so far which would provide the evidence of Altshuller's trends of the engineering system evolution. As it comes to the famous S-curve evolution trend, the "quality of the system" or "system performance," it is very easy to understand parameters for an informal talk, but almost impossible to agree on indicators for a quantitative assessment. We are not able to represent the evolution of a real engineering system by a single index. And the term "engineering system" requires an abstract level of analysis only. We should not immediately take a new idea of an engineering system in the form of a patent seriously, because many patents never become relevant, as some of them simply contradict the laws of physics. If it is new to a market system, what if it miserably fails as a product after a short period of time? Should we count lab prototypes or even gadgets that never became mass production? If not, what criteria can be applied for an engineering system to be legitimate as an event in relation to the S-curve analysis?

There are many more questions to be answered before a school of thoughts enters the level of scientific evidence. Otherwise it never leaves the domain of conventional wisdom, anecdotes, and rumor. For example, there is a famous number known to every TRIZnic: "40,000". Yes, this is the number of patents studied and analyzed by Altshuller to extract the TESEs and other TRIZ instruments (it means that TRIZ knowledge is a typical big data or literature-based discovery, performed manually). Altshuller reported he studied 40,000 patents. But the study was not documented in a way to be reproduced to become the basis for further development. We are not able to build this pool of 40,000 patents again, unfortunately, and this part of TRIZ became a part of literature, not science. The consequence is remarkable: the authors of scientific papers introduce the history of

TRIZ and have nothing but an anecdote to refer to. But the greater the number, the more impressive it is. Thus, we come across "100,000", "400,000 patents studied by Altshuller," and even "2 million patents TRIZ is based on," even in scientific papers.

Theories are to be scientifically proven but could it be true that the biggest theories do not need a proof?

Indeed, if many findings of Altshuller have been widely implemented in the practice of engineering conceptual design and if they inspired much scientific research (obviously, Altshuller is the most cited author in TRIZ-related publications), isn't it already beyond standard scientific contribution, which performance is measured by citations?

3 Academic Research on TRIZ

However, the fact that there had been no TRIZ-related publications in scientific journals until the late 1990s resulted in certain difficulties in TRIZ acceptance, deployment, and integration. It was rather risky to implement an approach that had never been acknowledged by science.

Fortunately, from the year 2000 onward, TRIZ received increased interest from those who prefer to publish research results in journals, indexed by leading scientific databases. In turn, these publications provide structured material for understanding TRIZ acceptance and development, bibliography analysis, trends of evolution, and open discussion. Thus, the past 15 years of evolution of TRIZ in scientific literature resulted in approximately 1000 peer-reviewed papers. It is a valuable material to understand how TRIZ is used and developed de facto. What are the most popular TRIZ tools and where are they typically applied? How is TRIZ being integrated into the roadmaps of modern engineering design? What are TRIZ competitors and what are the winning combinations with other design or research practices that promise high synergy? These and other questions are being discussed nowadays, which we deem to be a very good development.

Obviously, a review on scientific publications related to TRIZ deserves more attention than an editorial can provide. Moreover, a suitable review has recently been published (Chechurin 2016), and it is worth highlighting some results of this work: research efforts' distribution and noticeable trends.

The majority of TRIZ-related scientific contributions stay in the following paradigm: the theory is used for new product or technology design. Researchers either customize TRIZ tools slightly to fit certain application fields (e.g., chemical engineering or environmentally friendly design) or to demonstrate the power of the approach by design case studies.

An increasing share of studies uses TRIZ elements in an exciting hunt for successful "automated concept generation algorithms." The research question appears to be simple: can an algorithm provide a new idea? This is an interesting intersection of artificial intelligence, computational linguistics, and literature-based discovery where TRIZ "subject-object-action" and function analysis frameworks turned out to be a promising ontology. Other TRIZ tools like the contradiction analysis or trends of engineering system evolution support a field of research where huge amounts of texts (typically patents) are processed in order to retrieve "interesting" documents, to cluster them, or to distill certain trends and tendencies.

Worth mentioning is also a relatively small, but very high-cited share of publications, which use TRIZ for bridging between engineering and biology. Being one of the production samples, we readily assume that Mother Nature is a very successful designer, but the problem is that "The Designer" does not share the records. We do not know why some "designs" are so successful, but even when biologists discover the secret we need to database it in such a way that it is easy to access it with engineering domain requests. TRIZ turned out to provide elements of architecture for this database, for example, a function or contradiction-based phenomena description.

Finally, much effort is invested in applying TRIZ for nontechnical fields, like new service design, management, and business. For example, the inventive principles are either illustrated by the examples of smart managerial solutions or rewritten in the language of corresponding fields. Many authors present roadmaps for the integration of TRIZ in the product research and development process. In the same manner, researchers try to find a synergy between TRIZ and other more established methods for product design and development like OFD, Six Sigma, Lean, etc. The weakest points of these studies seem to be that proof is basically substituted by one or two case studies of design instead of empirical or statistical evidence.

TRIZ still seems to have been experiencing difficulties in enhancing idea generation in abstract fields, which deal with nonmaterial objects. For example, a negligible small amount of studies applies the theory for such a remarkable industry as coding, programming, or algorithm design. One reason could be that TRIZ is most effective in real, not abstract problems, where the thinking inertia originated by the conventional way of using certain material objects. TRIZ helps to focus on the functionality of the object, to substitute the material object by an abstract model in a similar manner as a mathematical model replaces the mechanical object in physics. But when the departure point is already nonmaterial, like an element of code, a big deal of TRIZ tricks does not work and even definitions become inapplicable. We are not able to define interactions, operation time, and an operation zone for software. Furthermore, ideality is to be redefined because the cost of material (the lines of code) is not going to be of much concern, the trend of evolution from mechanical structures to fields is inapplicable, etc.

Unfortunately, the typical TRIZ application paper engages contradiction analysis only. It creates the same distortion of TRIZ potential as if one claims that arithmetic is all in mathematics. The engineering contradiction elimination technique is simple and attractive to impress neophytes, but professional engineers would immediately reveal its weaknesses: the formulations of contradictions and inventive principles are very generic and do not differ much from brainstorming; they overlap and are nonuniform (compare inventive principle "use strong oxidants" and "change parameters").

Finally, before briefly introducing each contribution of this book, we present the statistical analysis which shows that "the amount of TRIZ research," measured by the amount of papers on the subject, is growing from less than 5 publications per year before 2000 to about 150 publications per year after 2012. The dataset was retrieved by the filtering publications with the word "TRIZ" in the Title, Abstract, or Keywords (TAK) fields. We could simply call it a "growing interest to the topic," but the total amount of related scientific papers in SCOPUS also shows similar growth. It is also worth mentioning that about 90 % of TRIZ-related scientific publications are hosted by the journals with very low visibility; the impact factor of these editions hardly exceeds 0.1. Only about 3 % of publications are made in journals with an impact factor exceeding 2.

We also notice that the "total amount of TRIZ research" measured by the total amount of publications (about 1200 by 2014) is comparable to the amount of studies which are related to practicing TRIZ techniques. The details are given in Table 1, which also shows the context of TRIZ in adjacent fields of knowledge.

4 Overview of Chapters

The departure point of *Elevate Design-to-Cost-Innovation Using TRIZ* by Zulhasni bin Abdul Rahim is the statement that "there is no specific tool that focused on solving cost problems explicitly" in TRIZ. However, Altshuller made this very clear in one of his book: cost is not the only engineering parameter; it is to be further expressed through technical parameters. In other words, we have to analyze why the cost is an issue. Potential questions might be is there labor-intensive manufacturing? Excessive use of expensive materials? The need for high-precision measuring? When the cost reduction is the primary goal of system redesign, TRIZ application yields ideas how to simplify the product of technology (see also DFMA rules). In general, simplification means fewer amounts of parts or technology operations that reasonably correlate with lower material or manufacturing costs. However, this does not imply that the efforts to link the function design with cost design should not be undertaken. The earlier the designer is able to see the economic projections of his/her design, the better. The study provides an illustrative mechanical design example showing how TRIZ application helped to reduce the costs dramatically.

Unfortunately, TRIZ was born and developed in a country where concerns about environmental protection were not among the highest priorities. Environmental issues are rarely discussed in TRIZ classics and not directly addressed by TRIZ instruments. For example, the Altshuller matrix does not bear such engineering parameters as the "harm for the environment" or "excessive pollution." They are to be generalized to "excessive use of energy," "substance loss," etc. Altshuller followers keep focusing on design for functionality or profit, unless the environmental problem appears in the context of chemical field or process control. In contrast, the share of scientific publications on applying/adapting TRIZ for eco-centered design is growing steadily. The study by Issac Lim *The Effectiveness*

Table 1 Context of TRIZ studies in indexed literature by July 2014 (Chechurin 2016)

	Total amount of papers with "*" in TAK fields, total amount	Column 2 selection AND "TRIZ" in TAK fields, total amount (relative amount)
"TRIZ"	1200	1200 (100)
"Computer-aided innovation"	93	56 (60)
"C-K theory" (design reasoning)	58	7 (12)
"Synectics"	40	4 (10)
"Axiomatic design"	740	51 (6.9)
"Kano model"	269	18 (6.7)
"DFSS"	400	15 (3.7)
"DFMA"	260	6 (2.3)
"Technology forecasting"	900	20 (2.2)
"Theory of constrains"	900	16 (1.8)
"Brainstorming"	2350	35 (1.5)
"Quality function deployment"	5100	74 (1.5)
"Six sigma"	4000	34 (0.9)
"Case-based reasoning"	7200	24 (0.3)
"Robust design"	3500	17 (0.4)
"Creativity"	31,600	130 (0.5)

of TRIZ Tools for Eco-Efficient Product Design is a nice example of it. It provides an overview of eco-related studies with TRIZ, statistical analysis of TRIZ tools applied for these problems, and an introduction of a new design tool, the ECO ideality chart. The tool application is illustrated by three examples.

Advanced Function Approach in Modern TRIZ by Oleg Feygenson and Naum Feygenson develops a function-based analysis. First, the study provides a nice introduction to the conventional function analysis that became a part of modern TRIZ and a popular analysis method. The authors highlight its weak points however. The latter is addressed by adding time and location variables. In a way it is the re-appreciation of classical TRIZ operation time and operation zone analysis tools that have been neglected in modern function analysis. A famous toothbrush benchmark example illustrates that the approach named "Advanced Function approach (AFA)" is capable to develop the picture of system functioning and differentiates the function performance in a more specific way. Another example of the simultaneous operation of two identical engineering systems shows that the new approach is capable of modeling the synergetic effect. *Using Enhanced Nested Function Models for Strategic Product Development* by Horst Nähler and Barbara Gronauer also adds to the function analysis technique. The study views the function model through the prism of the famous nine-screen vision of Altshuller. First it highlights the advantages of element nesting: a standard model transformation technique in system analysis (e.g., see IDEF0 technique for system hierarchy analysis or Simulink's "masking" option for system control circuits). In the same manner, it suggests to group/ungroup function model components in subassemblies. Secondly, it introduces the past, present, and future into a standard function model. A design case study illustrates the advantages of the suggested approach.

Interestingly enough, both studies focus on adding the time axis to the function modeling approach. It echoes the dynamic function modeling approach introduced in Chechurin et al. (2015).

Vladimir Petrov presents his original TRIZ-based algorithm for problem analysis in *5-Step Method for Conceptual Idea Design*. His TRIZ journey was initiated by G. Altshuller himself more than 40 years ago; Vladimir was his student and, further, active member of community of TRIZ developers. The enormous experience of TRIZ teaching and application resulted in the presented TRIZ tool application roadmap. Indeed, although ARIZ is still the one and the only sacred instruction of TRIZ tools' application in theory, the reviews show that the practice of ARIZ application is negligible. It is reported to be difficult, complex, and too demanding to learn.

Considering its name, TRIZ already bears one issue. The denomination "problem solving" seems rather ambitious and does not go along with the word "theory" very well. Imagine titles such as "theory of mechanical problem solving" or "theory of chemical problem solving." The main issue of TRIZ is the definition of the "problem" and finding a solution to the problem. Unfortunately, in contrast to mathematics, where the solution simply turns the equation into certainty or fact, the "solution" in TRIZ seems to be rather an optimistic substitute for a more appropriate "idea," "concept," or a "version" as far as design problems are concerned. TRIZ is an excellent ideation aid but it takes much more for an idea to become a real-world saving reality. With this philosophical tune, we consider the chapter *Taming Complex Problems by Systematic Innovation* by Claudia Hentschel and Alexander Czinki. It starts with a discussion on the basic definitions: problems, simple, chaotic, complex, and complicated problems and their place in innovation management. It is interesting to observe an attempt to interpret the concepts of nonlinear dynamics and system control for the much less formalized field of innovation management. The role of TRIZ in taming these problems is shown, although at a very generic level.

Another contribution on the same philosophical level is *TRIZ and Big Systems* by Dmitry Bakhturin. Here we face the definition of big systems as a big-scale business or company. The chapter speculates on the features of TRIZ deployment at the big company, for example, the necessity to consider man-machine systems, where classical TRIZ machine analysis-oriented tools may not work. The author highlights the difference between the canonized term "evolution trends" (in English), while Altshuller's original meaning in Russian was closer to "development trends." He also points out that the traditional model for a "supersystem" concept does not seem to be very productive when we deal with meta-systems in this context.

Since its first publication as a part of TRIZ, the trends for engineering system evolution (TESE) have been used to track and predict the evolution of artificial systems. But the numerous publications reveal an analysis performed on material products like airplanes or monitors. The *TRIZ-Evolutionary Approach: Main Points and Implementation* by Victor D. Berdonosov and Elena V. Redkolis is an innovative attempt to present the evolution in nonmaterial artificial systems: briefly in programming languages and more extended in numerical methods in mathematics. It is work of high interest: not much can be found in the literature regarding the application of TRIZ in programming, algorithm design, and, finally, mathematics. Indeed, most of the methods invented, even in such a logic-intensive science like mathematics, are the result of heuristic design. Since they are inventions, a natural question appears: could they be described by contradiction elimination, TESE, and other TRIZ instruments? The study provides an interesting classification to the huge family of numerical methods and a picture of their evolution.

TRIZ was born as the tool for engineers to design something new. Obviously, all the tools of this type are to be of interest for innovation managers and the researchers in the field. One question of these studies is where and how to integrate TRIZ with other tools in the innovation roadmap; another is how to apply TRIZ for innovation marketing and management directly. The chapter *Contradiction-Centred Identification of Search Fields and Development Directions* by Verena Pfeuffer and Bruno Scherb speculates on these two subjects and brings one more roadmap of TRIZ-assisted innovation.

TRIZ-Events Increase Innovative Strength of Lean Product Development Processes by Christian M. Thurnes, Frank Zeihsel, Boris Zlotin, and Alla Zusman provides one more TRIZ-assisted development process pattern. Classic and modern TRIZ tools are integrated into the lean-event roadmap. The study speaks the language of an international ideation company, which develops their own methods and products for invention support: Innovation Situation Questionnaire (ISQ), Anticipatory Failure Determination (AFD), direction for innovation, Direct Evolution, and Source-Effect-Object-Result Model (SEOR)), among others.

The next part of this book is the collection of case studies. *TRIZ in Enhancing of Design Creativity: A Case Study from Singapore* by Iouri Belski, Teng Tat Chong, Anne Belski, and Richard Kwok open that part with a model case. It reveals a documented mechanical design improvement process assisted by TRIZ. The results are patented and implemented—what could be better as a success story?

Another illustration is *TRIZ-Supported Development of an Allocation System for Sheet Metal Processing. A One-Day Case Study* by Barbara Gronauer and Horst T. Nähler. The report contains a documented case of a TRIZ-guided brainstorming session of a team of engineers that lead to a "qualified, capable solution concept" in redesigning an existing machine.

TRIZ as a Primary Tool for Biomimetics by Julian Vincent opens the part of free essays. It is a pleasure to have a chance to host the author of most cited TRIZ-related publications in this book. G. Altshuller wrote in 1961 "Unfortunately, inventors cannot easily use the "patent database" of Nature. Engineering knowledge is not yet linked to the biological one." Addressing this point, the chapter reviews the advance of biomimetics and the role of TRIZ in making the technology transfer from nature to engineering more systematic.

In contrast to what is stated in the beginning of *Using TRIZ in the Social Sciences: Possibilities and Limitations* by Joris Schut, there is actually a big amount of studies on adapting/applying the use of TRIZ in nonengineering fields. The essay meditates on the subject at a very general level and provides a reasonable conclusion that states that more work needs to be done to adapt TRIZ for social sciences.

Linking TRIZ and Cross-Industry Innovation—Evidence from Practice. How TRIZ in the Context of Cross-Industry-Innovation Can Turbo-Charge the Innovation Process by Peter Meckler is the interesting free speech text based on the experience of an innovation facilitator. It tells how TRIZ was used in many projects in multi-field engineering teams to support the ideation stage. TRIZ (or what the author believes to be TRIZ) is placed among other creativity methods in a nonsystematic way. This text is vivid reading with insights and humorous anecdotes.

Finally, the *Glossary* by Valeri Souchkov is believed to be a useful reference for TRIZ terminology used in this book and outside of it.

To conclude the editorial before we briefly introduce the chapters of our book, we anxiously predict that TRIZ has a challenging but bright future in the domain of science. It might undergo some critical revisions and transformations, get rid of personal and historical influence, doubtful, biases, and unnecessary pieces, and even fall apart into several elements. But these elements can become the cornerstones for the further systematization of heuristic acts, hypothesis construction, and ideation.

Acknowledgments I would like to acknowledge TEKES, the Finnish funding agency for innovation and its Finnish Distinguished Professor (FiDiPro) program that supported the research.

References

Altshuller, G. S., & Shapiro, R. B. (1956). Psychology of inventive creativity. Vopr. Psikhologii (Issues Psychology), no. 6.

Chechurin, L. S. (2016). TRIZ in science. Reviewing indexed publications. *Procedia CIRP, 39*, 156–165.

Chechurin, L. S., Wits, W. W., Bakker, H. M., & Vaneker, T. H. J. (2015). Introducing trimming and function ranking to solid works based on function analysis. *Procedia Engineering, 131*, 184–193.

Devoino, I. G., Koshevoy, O. E., Litvin, S. S., & Tsourikov, V. (2011). Computer based system for imaging and analyzing a process system and indicating values of specific design changes. US 6202043 B1.

Tsourikov, V. M., Batchilo, L. S., & Sovpel, I. V. (2000). Document semantic analysis/selection with knowledge creativity capability utilizing object (SAO) structures. US6167370.

Part I
Scientific Articles

Elevate Design-to-Cost Innovation Using TRIZ

Zulhasni bin Abdul Rahim and Nooh Abu Bakar

Abstract Design-to-cost (DTC) is a powerful concept to adopt in reducing cost at design level. The concept brings the cost parameter to the same level with the design or technical parameter. The ultimate goal of DTC is to design a product that effectively meets the planned target cost before the product is launched. Therefore, DTC consists of tools which assist the organization to achieve its goals. However, the effectiveness in achieving its goals is quite challenging as there are a number of conflicting issues in the process of driving down the cost toward the target cost. The best and most common tool of the DTC concept is a trade-off. A trade-off allows designers to tune their designs and seek ultimate points of optimization between conflicting product requirements. This directly hinders the designer from pushing the cost further down or achieving the targeted cost as it is only looking for a compromise as its solution. A framework called design-to-cost innovation (DTCI) is introduced to overcome these challenges. A case study is shared to discuss the application of the DTCI framework as compared to the optimization approach. The application of TRIZ tools in DTCI managed to achieve 75.3 % in weight reduction as compared to 22.1 % from the optimization approach, which indirectly reduces the material cost of the system. The outcome of DTCI brings a higher value to cost reduction initiatives by eliminating trade-offs and improving product innovation.

Keywords TRIZ • Design-to-cost • Cost reduction • Optimization • Automotive

1 DTC and Its Constraints

The first DTC concept was introduced in the military industry by The Department of Defense (DoD), United States of America. The concept was applied through DoD Directive 5000.1 named "Acquisition of Major Defense Systems" way back in 1971. The initial objective of this directive was to quantify the design parameter in the form of cost parameter. This established the cost element from the design

Z.b.A. Rahim (✉) • N.A. Bakar
UTM Razak School of Engineering and Advanced Technology, UTM Kuala Lumpur, Jalan Sultan Yahya Petra, 54100 Kuala Lumpur, Malaysia
e-mail: zulhasni@gmail.com

L. Chechurin (ed.), *Research and Practice on the Theory of Inventive Problem Solving (TRIZ)*, DOI 10.1007/978-3-319-31782-3_2

parameter which gave impact to development cost and product cost. Later, another directive was created, DoD Directive 5000.28 named "Design-to-Cost" to improve the adoption of new concepts as guidelines which later become a policy. The most significant change in the new directive was highlighting cost control toward preestablished target cost throughout the design and development process of the product. At that time, the only approach which supported the product developer to achieve the given target cost was by adopting a trade-off between cost and other critical deliverables such as product performance, product design parameters, development time, or product quality.

The practical trade-off approach adopted by the DoD was considered as the most feasible method to achieve the target cost which was focused on finding a balance point between conflicting goals in product design and development (Tyson 1989). In other words, practical trade-offs would seek a compromise between the product design parameter and product cost parameter to prevent the final cost of the product to go beyond the targeted cost (Rahim and Bakar 2013). This forced the DoD to explore more effective methods or tools for support to achieve the target cost. Furthermore, they needed a tool which provided a specific analysis on the design and cost parameters in order to assist them in controlling the product cost from going beyond the target cost and eventually fail the project (Montgomery and Carlson 2011).

Subsequently, value engineering (VE) was adopted as a tool to reduce the dependency on trade-offs by analyzing between design and cost parameters. Wichita (1975) stated that VE was able to provide a significant improvement to DTC by incorporating clauses in the project's contract. Wichita (1980) conducted several case studies on the application of VE in DTC projects to develop weapon systems, which in his opinion was successful. However, the study recommended that trade-offs were still a component of DTC projects followed by the VE method to achieve target cost (Zulhasni and Nooh 2015).

The vertical improvement of DTC effectiveness to achieve target cost was not merely by introducing VE into the processes. Several tools have been proposed throughout the four phases of DTC based on a comprehensive DTC framework by Gilb and Maier (2005). The DTC framework comprises the following phases in sequence: preparation, design, evaluation, and implementation. Figure 1 shows the tools proposed in the DTC processes based on the framework by Gilb and Maier.

A common tool used in the preparation phase is the Pareto analysis, which focuses on prioritizing improvement areas for DTC projects. In the design phase, tools such as VE analysis and brainstorming are used to generate ideas to achieve the target cost. In the evaluation phase, the DTC project would encounter problems which may become constraints to its goals. Common problem-solving tools are used in this phase, such as "5-Why analysis" (Gilb 2011). However, there is less options for the DTC project in solving problems as it marches toward the implementation phase. In this phase, there is only one common alternative left for the DTC to execute the project, which is using the trade-off analysis. This tool distinctly proposes a compromise between conflicting needs, especially in terms of the cost parameter (Williamson 1994).

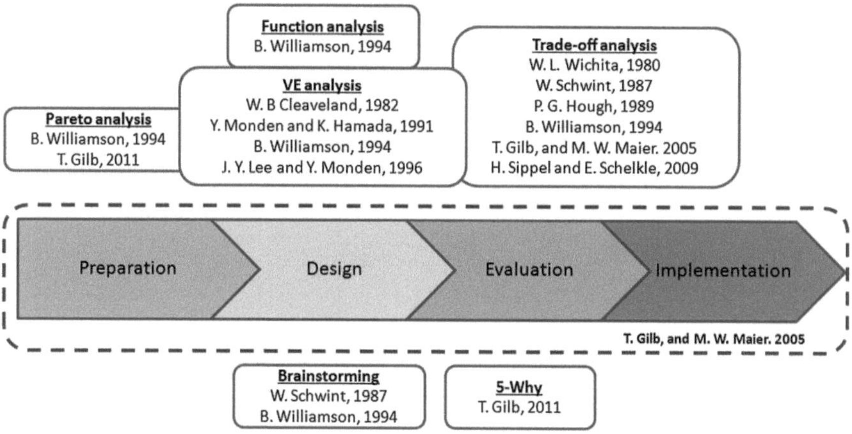

Fig. 1 Application of tools in DTC processes based from Gilb and Maier's framework

There is also a horizontal improvement that is focused on creating better value compared to the DTC. A new concept called "Cost as an Independent Variable" (CAIV) was introduced to the DoD in 1995. The CAIV concept highlights cost as a fixed variable, while performance and schedule are allowed to vary (Boudreau 2006). In other words, the focus on compromise is transferred to performance and schedule. Meanwhile expecting the product of the project is affordable. However, this concept is not feasible when the project is extended to a longer schedule. This is because the operational cost is still active and therefore, the total development cost would increase. It would have a similar impact on the compromising performance to achieve target cost, which inevitably ends up with poor customer satisfaction (Zulhasni et al. 2015).

The vertical and horizontal improvements of DTC are still tied to the trade-offs as their final decision-making in pursuing the target cost. However, in 2000, Esaki claimed making the first attempt to introduce TRIZ in DTC, together with other concepts such as quality function deployment (QFD) and the Taguchi method (Esaki 2005). Figure 2 shows the evolution of the DTC concept derived by Esaki (2005).

The main possible reason for the new method such as TRIZ to become a part of the DTC concept is to overcome the dependency of trade-offs in the main processes. This opens up a new improvement in the overall DTC concept if TRIZ is to be significant to break away from trade-offs. However, the search for literature on a proposed framework(s) and case studies on implementing TRIZ in DTC has yet to be found (Bakar and Rahim 2014). This creates motivation to pursue the possibility of the DTC in adopting TRIZ in its framework and processes. The next section will discuss some investigations conducted on how TRIZ was adopted in cost reduction initiatives, which was similar to the DTC concept. Subsequently, TRIZ was applied in the DTC concept through several case studies using a new framework called design-to-cost innovation (DTCI).

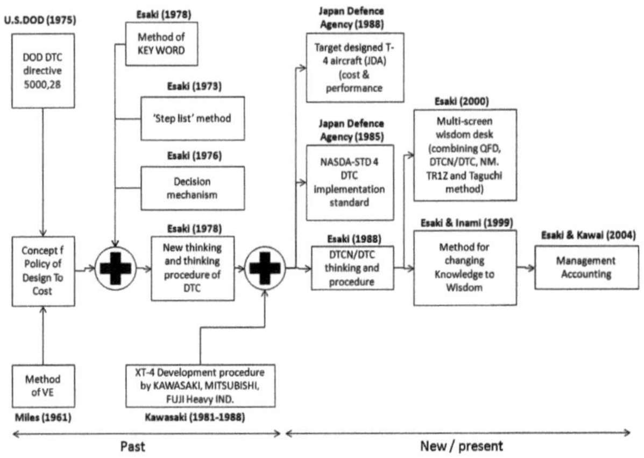

Fig. 2 The evolution of DTC concept (Esaki 2005)

2 TRIZ in Cost Reduction

Most TRIZ practitioners are aware and may agree that TRIZ is an arch enemy of trade-offs (Hipple 2012; Linstone 2011). One of the main reasons for the existence of the TRIZ methodology is to break away from the compromise or trade-off. Without TRIZ, most people would do their best and focus on optimization until they reach the ultimate limit (Hipple 2012; Cascini et al. 2011). Most probably people at this stage are unable to think of better solutions and instead propose more complex solutions (Rahim and Nooh 2014). Furthermore, what they require deploys unnecessary resources to maintain high levels of optimization. This includes cost as one of the main bottom-line for any industry.

Cost is the problem of all industries and things get more severe when the competitive environment becomes hostile. Almost all industries are struggling to improve their cost at every level of the business process. Regardless of how cost reduction is done, most business owners only want to see huge profits and zero losses. They would use whatever methods or approaches to find the ultimate solution reduce cost, including employing TRIZ methodology (Sheu and Hou 2011).

Many tools from level 1 to level 3 are taught within the scope of knowledge governed by the International TRIZ Association, or known as MATRIZ. The purpose to divide to three levels is to ensure that TRIZ practitioner is able to adopt the complexity of the methodology. Level 1 tools are "function analysis," "cause effect chain analysis," "ideality," "trimming," "engineering contradiction," "contradiction matrix," and "40 inventive principles." In level 2, the tools are "physical contradiction," "Su-field analysis," "76 standard inventive solution," and "S curve analysis." The rest of the 14 more tools are allocated in level 3 that are mostly known as modern TRIZ tools. However, there is no specific tool that focused on solving cost problems explicitly (Stratton and Mann 2003). Most of the applications of those tools are used to indirectly reduce cost-related problems. For example, in the application of Contradiction Matrix, there is no "Cost" listed as worsening or improving parameters, neither is "Cost" listed as a part of inventive principles (Domb 2005).

Cost is considered as subjective and is dependent on the context of the moment. In cost reduction initiatives, there are many conflicting factors caused by identified cost element(s). This cost element may have a direct, inverted, or exponential relationship with the technical parameters of TRIZ. Furthermore, some inventive principles are capable of providing effective solutions while some do not. For example, in the context of meeting customer level of affordability, segmentation of product variants may solve the problem. However, merging many variants of a product may reduce the number of its resources, which could impact on cost reduction. So, which solution is better? Create segmentation to expand the market or merge to reduce resources. This, of course, creates another contradiction to solve.

Furthermore, the buzzword in the twentieth century competitive industry is innovation. The industry is pushing new technology to the market; at the same time the market is pulled by customer demands for better products from the industry. This automatically imposes a greater challenge to the industry to ensure that they survive in the competition. There is a misconception by industries regarding innovation that it always requires a huge investment and incurs great risk to the organizational performance. This hinders industries from pursuing innovation, and instead they choose to conduct business as usual, hoping that they would survive any competition coming their way.

In investigating how other TRIZ practitioners carry out cost improvement activities, a literature review was conducted on areas of cost reduction. Most TRIZ practitioners are focused on the product design area as it brings a huge impact and delivers significant results in cost reduction. Domb and Kling (2006) suggested that the focus on cost improvement must begin from the cost of the root cause(s). Domb (2005) recommended several TRIZ tools as in Fig. 3 which was considered as effective to solve cost problems.

Isaka (2012) proposed cost cutting in redesigning products to develop simpler products using "Trimming" in the 8th TRIZ Symposium in Japan. The trimmed system was expected to create new problems to be solved to achieve cost improvement results. Furthermore the focus of TRIZ tools in cost reduction initiatives has expanded to other improvement concepts. Ikovenko and Bradley (2003) advanced an integrated TRIZ under Lean Thinking Tools in the 2004 ETRIA Future Conference. The objective of integration was to harness the advantages and potential which could be effectively used in organizational methods like Lean. The strongest TRIZ tools applied in Lean were "Trimming" and "Flow Analysis."

Besides conceptual studies, there were also several case studies which used TRIZ in cost reduction initiatives. Cho et al. (2004) shared their application of TRIZ in reducing material cost in a Samsung camcorder product. The main TRIZ tools used in their product were Function Analysis, Technical Contradiction, and Su-Field Analysis. The most interesting outcome of the cost reduction activities was their success in securing three new patents for the technological innovation.

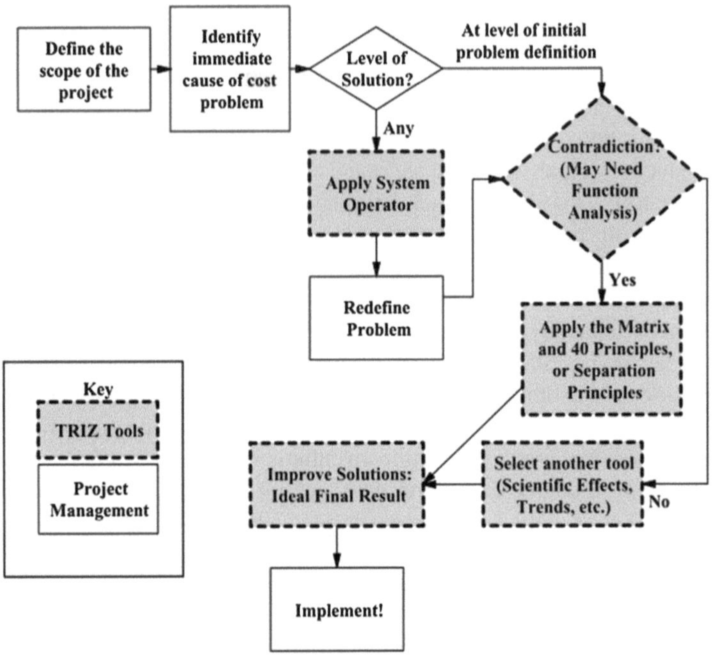

Fig. 3 Flowchart for cost problem using TRIZ (Domb 2005)

There were some studies on the application of TRIZ tools which provided significant improvement to the current DTC which moved away from using trade-offs. Table 1 shows a summary of previous research on TRIZ applications in cost reduction and area of complement in the DTC process.

The common tools used and the approach adopted by TRIZ practitioners in cost reduction can be seen in Table 1. One of the strategic approaches is by integrating TRIZ tools into other concepts. Each context of TRIZ tool application varies depending on the objective of the cost reduction initiative. The following section describes a new framework of DTC which integrates TRIZ tools in order to achieve better cost reduction performance without trade-offs and also improves the level of innovation.

3 DTCI Framework

A framework is created based on the inherent contradiction of current DTC in product design and innovation. A new version of the DTC framework called DTCI framework was developed using the framework by Gilb and Maier (2005) as a base. The framework consisted of four phases: system prioritization, idea generation, idea evaluation, and implementation. The unique part of integrating the TRIZ

Table 1 A summary of researches on TRIZ related to cost improvement

No.	Authors and year of publication	DTC area of improvement	TRIZ tools	Integration of TRIZ
1	Ikovenko and Bradley (2003)	Idea generation	Trimming	Integration success
2	Mann and Domb (2003)	Trade-off and evaluation	Contradiction	Harmful function
3	Sawaguchi (2000)	Brainstorming	Inventive principles	High-value ideas
4	Stratton and Mann (2000)	Idea generation and evaluation	Trimming	Engineering needs
5	Domb and Kling (2006)	Technical cost element	Contradiction	Integration
6	Domb (2005)	Prioritization	Inventive principles	Manufacturing success
7	Isaka (2012)	Design simplification	Trimming	Simplification
8	Mann (2002)	Idea generation	Function analysis	Business area
9	Martin (2010)	Prioritization	Trimming	Lean and TOC
10	Wu (2004)	Design improvement	Inventive principles	Taguchi method
11	Li (2010)	Trade-off	Contradiction	Technology acceleration
12	Mann and Domb (1999)	Design improvement	Function analysis	Customization
13	Ball (2014)	Simplification	Trimming	N/A

method is that it is divided into three categories. The first category is addressing the cost problem of the existing system. For example, components caused high defects based on warranty claims. The second category is developing a new system to improve the cost of the old system, due to changes in a stakeholder's requirements or meeting the target cost. The third category is developing an advanced system which improves the level of product innovation or which is related to developing patents. Figure 4 shows the DTCI framework with an integration of TRIZ tools.

All DTCI initiatives are strategically prioritized based on the product or components of cost analysis. This focus is on a specific system which applies TRIZ tools for improvement. The solution developed is evaluated from three dimensions: technical, commercial, and management. Solutions which meet the requirements of the evaluation process would proceed to the implementation process. Meanwhile, solutions which do not meet the requirements would either be improved further or reserved for future strategic product development.

There are recommended application tools in the DTCI framework within those categories, as shown in Fig. 5. Certain tools are suitable for a specific objective and context in solving a cost problem on an existing system, developing a new system, or exploring an advanced system. However, this does not restrict the application of other tools in other categories that is not listed as recommended tools. There are also other new TRIZ tools used in the DTCI framework which are not mentioned in

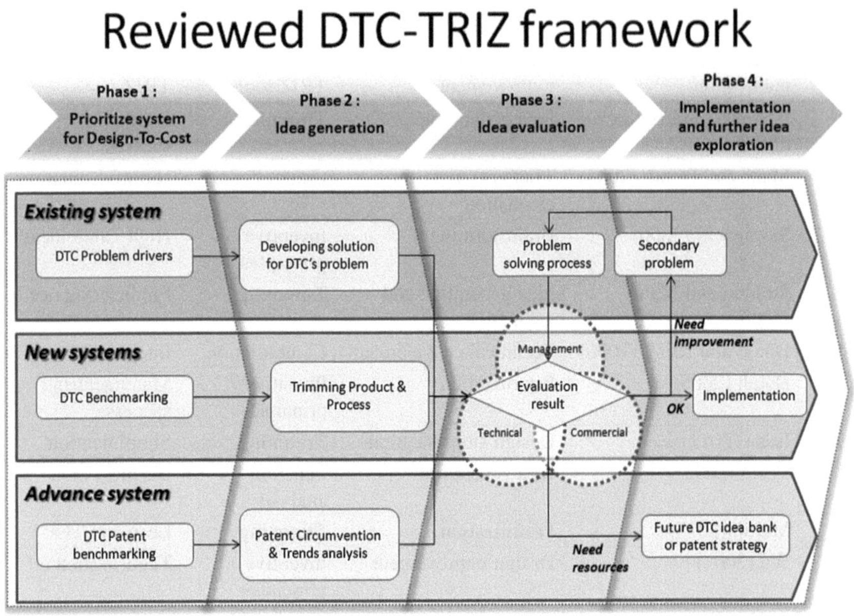

Fig. 4 DTCI framework

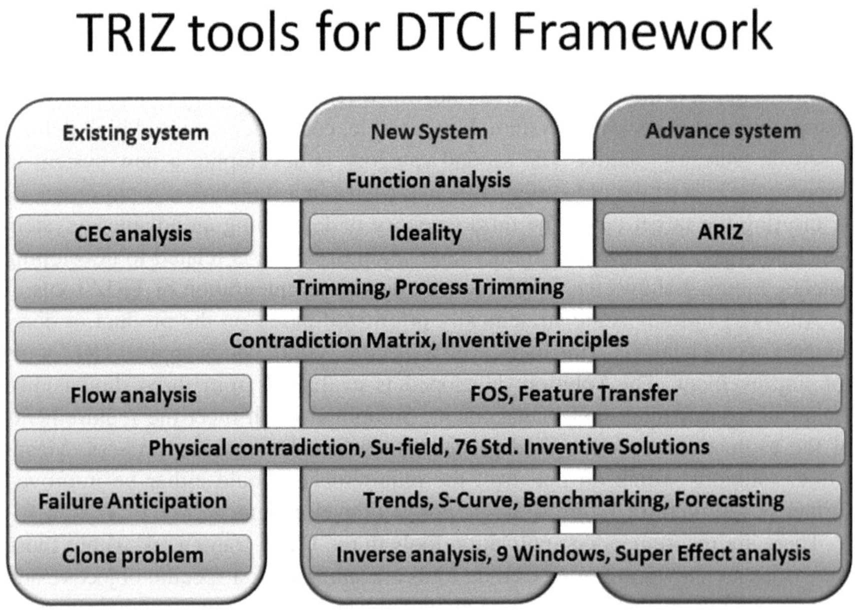

Fig. 5 Recommendation list of TRIZ tools in DTCI categories

the list such as patent circumvention and patent strategy, which consist of a combination of other TRIZ tools with a legal perspective.

The next section looks at some case studies on each DTCI category and the focus to enhance DTC to achieve better target cost and to improve the level of product innovation.

4 DTCI Case Studies

The DTCI framework is an elevation of the current DTC with the application of TRIZ tools and focuses to break away from being too dependent on trade-offs. Another expected outcome of DTCI is to embrace innovation in product design so as to achieve the cost reduction objective. There are several published projects concerning case studies on the application of DTCI in the automotive industry at the systems and component level (Rahim et al. 2015). However in this chapter, a case from one of the projects is presented on problem solving by way of a comparison study between a common optimization method and the TRIZ method.

4.1 Project Case: DTCI in Existing Design Optimization

A wiper system is selected for improvement through a cost engineering analysis. The wiper system has a number of limitations such as the issue of reliability and material cost. On top of this, the engineering team was intent on improving the design of the wiper system and reducing the weight of its components. Figure 6 shows the wiper system mounted on a car.

A cross function team was assembled. It consisted of members from engineering design, procurement, quality, and manufacturing including suppliers. The activity started with a component analysis, before it progressed to developing an improved design concept through current engineering optimization activity. After that the same system was analyzed using TRIZ methods and another design concept was developed for a comparison study. The focus outcome from both the design concepts was a comparison on component cost improvement, functionality, and key focus on reducing the weight of the wiper system.

The heaviest component in the wiper system was the wiper bracket. It weighed about 1.5 kg, which was considered too heavy for the total system. It was made from thick cast iron and went through the casting process to obtain its solid and rigid features. The features were required as this component was considered as the most critical component in the wiper system mechanism. If the weight of the bracket was reduced, it would directly reduce the material cost. However, the weight reduction must not reduce the reliability and performance of the wiper system. This was considered as a good problem statement for TRIZ to solve the

Fig. 6 Original wiper system in the current model

contradiction. Meanwhile, conventional methods began to seek trade-off points between weight reduction, cost reduction, and performance through design optimization.

In a product design optimization process, computer-aided engineering (CAE) would be used to analyze the current wiper system for weight improvement. Figure 7 shows the overview process of design optimization for the design concept of the wiper bracket.

Based on the optimization process using CAE, some minor changes were proposed for the existing design of the wiper bracket. The limitation to the changes was the allowable maximum stress applied on the bracket during operational mode. The reduction in weight was achieved by 22.1 %, which was equivalent to 0.332 kg from its current weight. Figure 8 shows the changes performed through CAE optimization.

The next method was using TRIZ as a weight optimization initiative. First, the system was modeled using function analysis to investigate the level of function which existed in the current wiper system. Figure 9 shows the function analysis performed on the wiper system. In the context of weight reduction, the wiper bracket carried excessive functions in holding other components. From an engineer's viewpoint, the excessive weight of the wiper bracket was to attain rigidity in holding other components and to have it mounted on to the body structure. This

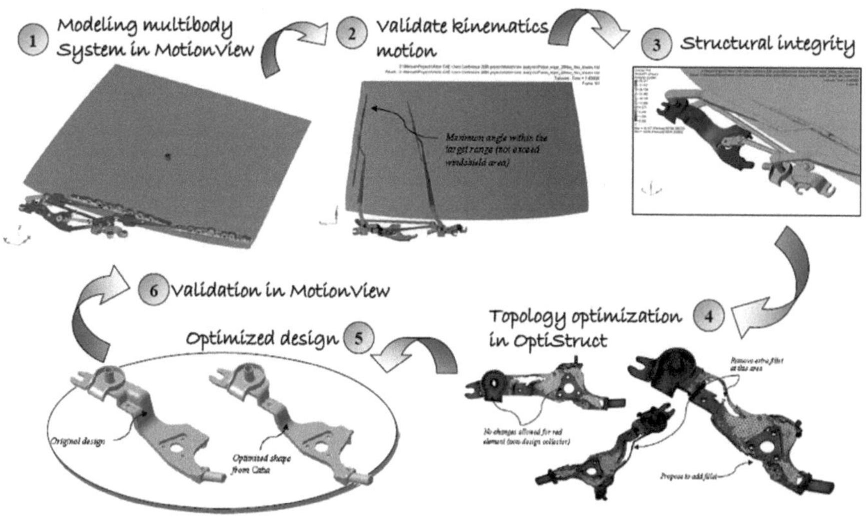

Fig. 7 Optimization process on weight reduction for wiper bracket

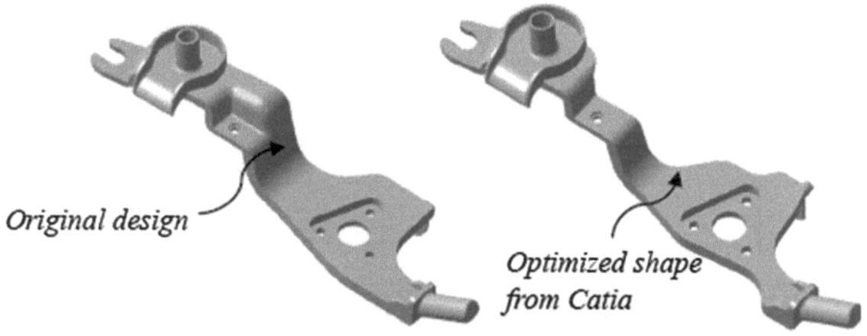

Design	Mass (kg)	% reduction
Original	1.504	-
Optimized	1.172	***22.1%***

Fig. 8 Comparison of optimization analysis and weight improvement

understanding is considered as psychological inertia to overcome the contradiction between weight and rigidity.

The contradiction was reviewed through 39 engineering parameters. The improving parameter comprised weights of stationary object (#2), whereby the stationary wiper bracket was fixed on to the body structure. Meanwhile, the worsening parameters were force (#10), stress (#11), stability (#13), strength (#14), reliability (#27), and ease to manufacture (#32). The worsening parameters were highlighted by the subject matter expert (SME) in the cross function team. Figure 10 shows inventive principles extracted from the contradiction matrix.

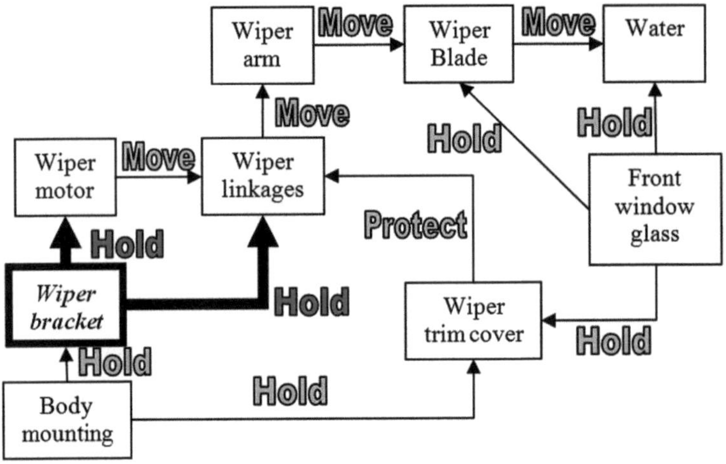

Fig. 9 Function analysis of wiper system

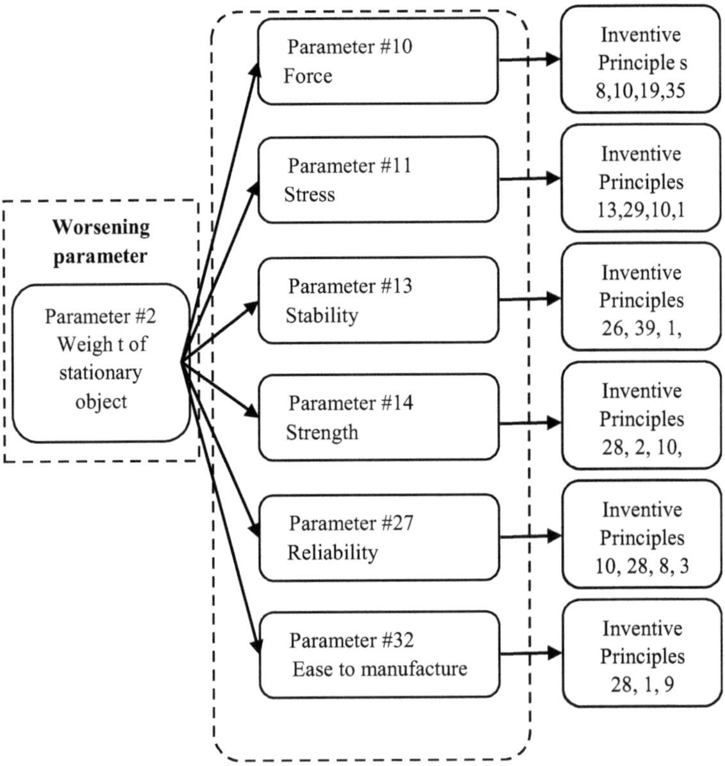

Fig. 10 Contradiction analysis between the studied parameters for weight improvement

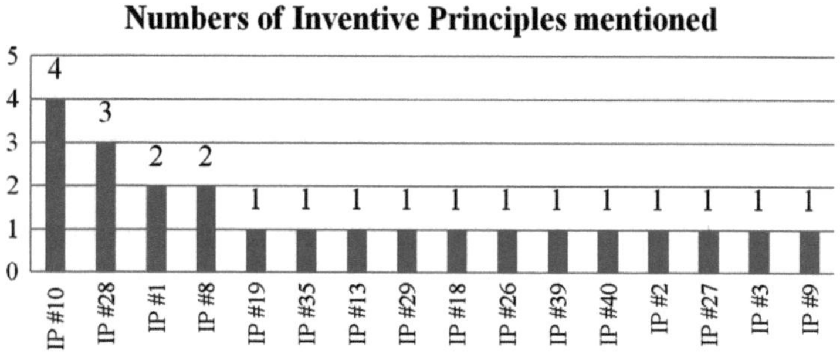

Fig. 11 Pareto analysis of recommended inventive principles for the optimized weight improvement product

The next process was to identify the most common inventive principles recommended to solve interrelated contradictions in reducing the weight of the wiper bracket. Using a simple Pareto analysis, the inventive principles were analyzed to identify the most common concept proposed with regard to similar contradictions. There were four inventive principles that were mentioned more than once. Figure 11 shows the four inventive principles: "Preliminary action," "Mechanics substitution," "Segmentation," and "Anti-weight."

Based on recommendations from many inventive principles, the cross function team was excited to explore all the inventive principles to develop the concept solution. The concept solution was focused on the mechanism of how the function delivered and the technical feasibility of the established concept solution. Table 2 shows the concept solution discussed based on recommended inventive principles and the results of technical evaluation. The technical evaluation result is based on the capability to develop the proposal using existing available resources determine by the DTCI team, experts, and project manager.

The cross function team used their current knowledge and experience to develop the wiper system through the concept of inventive principles. Quite a number of amazing concepts were generated and each of them required some evaluation study on its technical feasibility. However, the development of concepts was quite straightforward due to the team members' basic level of TRIZ knowledge and application. Some members utilized the Internet to extract information to support their proposed concept with facts and figures.

The evaluation process in DTCI also included a commercial study. The concept solutions were mapped to identify the most significant for design improvement using four-by-four matrix cost analysis and technical feasibility; Fig. 12 shows the commercial–technical matrix on the proposed concept solution. The solutions are allocated on the investment cost required and the feasibility of existing resources confirm by the DTCI team and manager.

The most feasible concept solution used the inventive principles: #1 (Segmentation), #35 (Parameter change), #8 (Anti-weight), and #10 (Preliminary action). The activity had made each team member more receptive to developing practical

Table 2 Ideas generated based on inventive principle into the optimization of the wiper system

Inventive principles	Specific improvement	Impact	Technical evaluation results
IP #10 Preliminary action	Use water repellent for front glass window	Not suitable for other than water	Feasible (moderate innovation)
IP #28 Mechanic substitution	High-frequency ultrasound in windshield	Able to delete existing components such as wiper motor	Feasible (radical innovation)
IP #1 Segmentation	Separate the bracket into two pieces	Smaller size bracket	Feasible (minor modification)
IP #8 Anti-weight	A portion of bracket is designed into the trim cover	Reduce metal material	Feasible (minor modification)
IP #19 Periodic action	Pressured air blows the water periodically	Reduce the force to wipe the water, reduce the bracket strength	Feasible (moderate innovation)
IP #35 Parameter change	Use hollow parts to substitute solid parts	Reduce the total weight of parts	Feasible (minor modification)
IP #13 The other way around	Change the position of wiper system to the top of the front glass window	Reduce the force to wipe the water, reduce the bracket strength	Feasible (moderate innovation)
IP #29 Pneumatic and hydraulic	Sucks in water from the glass window Similar solution as IP #19	Eliminate wiper system Similar solution as IP #19	Feasible (radical innovation)
IP #18 Mechanical vibration	Ultrasonic cleaning Same as IP #28	Eliminate wiper system Similar solution as IP #28	Feasible (radical innovation)
IP #26 Copying	Use bracket made of lightweight polymer	Need technical validation and testing	Feasible (minor modification)
IP #39 Inert atmosphere	Introduce a system that changes water into gas	Eliminate wiper system	Feasible (radical innovation)
IP #40 Composite material	Use composite material on metal and plastic parts	Need technical validation and testing	Feasible (minor modification)
IP #2 Taking out	Take out the bracket and use the body structure as support	Eliminate bracket and wiper needs to work Harmonically	Feasible (moderate innovation)
IP #27 Cheap short living object	Not available	The bracket cannot be used within a short period of time	Not feasible
IP #3 Local quality	Make the bracket/body structure part of the linkages to move wiper or Similar with IP #2	Need technical validation and testing	Feasible (moderate innovation)
IP #9 Preliminary anti-action	Glass surface filled with water	Further mechanism to sustain the water on top of the glass	Feasible (radical innovation)

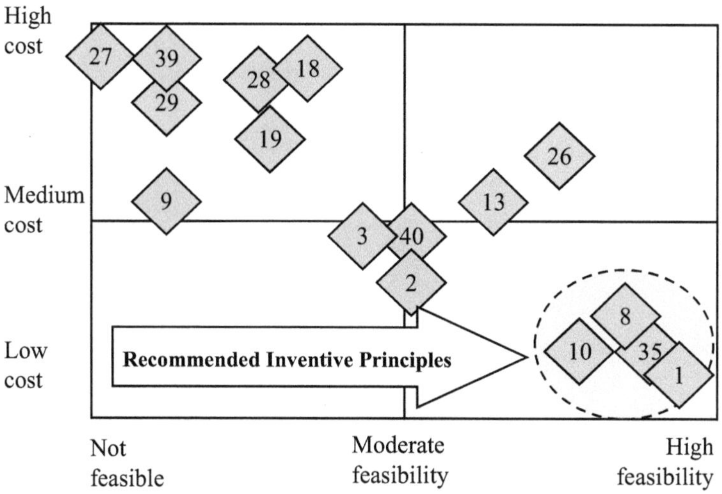

Fig. 12 Recommended inventive principles through technical and commercial evaluations

concept solutions which provided significant impact to their respective scope of work. The wiper bracket was changed to a new design, focused on better weight reduction and lower cost without compromising on its performance. However, the design engineer needed to consolidate and refine the concept into a real practical product which met the target cost and brought new innovation into the wiper system. Other concepts which were considered less feasible to implement due to time taken for development or exceeding the target cost were kept in a database for innovation research and development.

The wiper bracket was segmented into three parts. The middle part which was bigger in size was substituted with a hollow shaft. The rest of the part was modified to integrate with the hollow shaft. Another moving wiper bracket was also redesigned based on a similar concept. Plastic material was introduced into the new system in less critical functions, and metal bush was substituted with a special polymer bush which had better performance and reliability. The rest of the components were transferred to the new system such as the wiper motor, which enabled the cost to be reduced further. The final concept was developed at the prototype level and underwent testing and validation of part performance. Figure 13 shows the existing wiper system and Fig. 14 shows the new wiper system.

The results from this approach achieved 75.3 % in weight reduction, which was equivalent to 1.13 kg from the current design. The result of the weight reduction using TRIZ methodology produced a better outcome compared to the optimization approach which was supported by CAE software. Table 3 shows the weight of the wiper bracket for the existing and the new improved design. The improvement of the wiper system would continue as there were many solutions generated by the cross function team using TRIZ which require a longer time and adequate resources to move forward in product innovation.

Fig. 13 The current wiper link system

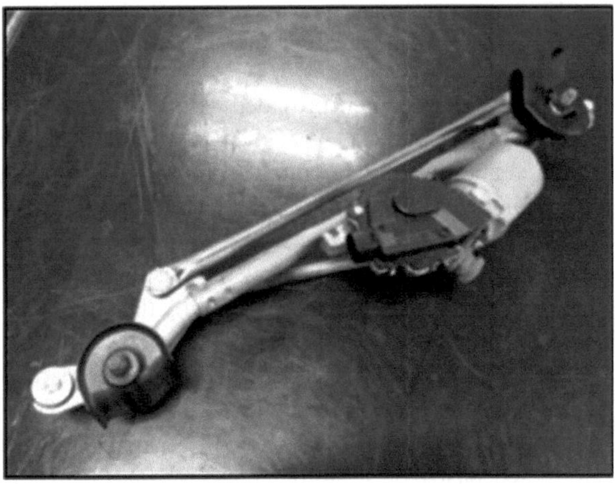

Fig. 14 New wiper link system with minor design modification

Table 3 Weight optimization and TRIZ design solution

Level of bracket design	Total weight (kg)	% of weight reduction
Current design	1.504	–
Optimized design	1.172	22.1 %
Adopting TRIZ inventive solution design	0.371	75.3 %

5 Conclusion

The main objective of DTC is to focus on achieving a planned target cost at the initial stage of design. One of the critical steps to achieve effective DTC is to translate a design or technical parameter to a cost parameter. However, in the process of achieving the target cost, there are quite a number of conflicting requirements such as bringing the cost down to a minimum without compromising on the design performance of the product. Currently, there are many methods and improvement concepts used in DTC phases, such as prioritization, idea generation, idea evaluation, and implementation. Among these tools, trade-off will be the most important tool used in DTC to achieve the optimum level of cost reduction against other parameters such as product performance. This chapter has presented a DTCI framework which integrated TRIZ in DTC processes and broke away from being highly dependent on trade-offs. A project case study was discussed. It addressed the cost problem by way of material weight reduction on an existing system. The outcome showed that TRIZ has elevated DTC by producing better value in solving DTC's problem compared to the trade-off method such as optimization. Furthermore, TRIZ has enabled the cross function team of DTC to enhance their product innovation for future development from an existing system, a new system, or an advanced system. The DTCI framework is expected to solve the contradiction between achieving the target cost and enhancing the level of innovation.

References

Bakar, N. A., & Rahim, Z. A. (2014). Design-to-cost framework in product design using inventive problem solving technique (TRIZ). *Journal on Innovation and Sustainability, 5*(2), 3–17. RISUS ISSN 2179-3565.

Ball, L. (2014). *TRIZ power tools: Simplifying* (2014 ed.). http://www.opensourcetriz.com/attachments/Simplifying.pdf

Boudreau, M. (2006). *Using cost as an independent variable (CAIV) to reduce total ownership cost*. Monterey, CA: Naval Sea Systems Command and Naval Postgraduate School.

Cascini, G., Rissone, P., Rotni, F., & Russon, D. (2011). Systematic design through the integration of TRIZ and optimization tools. *Proceeding of the ETRIA World TRIZ FUTURE Conference, Procedia Engineering, 9*, 647–679.

Cho, Y. H., Kim, J. S., & Lee, J. Y. (2004). *Innovation in the material cost of the 8mm camcorder using TRIZ* (pp. 1–17). TRIZCON 2004 Conference Proceedings.

DOD Directive #5000.1. (1971). *Acquisition of major defense systems*. Retrieved 17 August, 2015, from http://www.dtic.mil/whs/directives/corres/pdf/500001p.pdf

DOD Directive #5000.28. (1975). *Design to cost*. Retrieved 17 August, 2015, from http://www.valueeng.org/knowledge_bank/attachments/Application%20of%20VE%20to%20DTC.pdf

Domb, E. (2005, November). *How to deal with cost-related issues in TRIZ*. European TRIZ Association TRIZ Futures Conference 2005, Graz, Austria.

Domb, E., & Kling, T. J. (2006, April). How to reduce cost in product and process using TRIZ. *TRIZ Journal*, Issue. Retrieved 17 August, 2015, from http://www.triz-journal.com/deal-cost-related-issues-triz/

Esaki, M. (2005). *History of methodology of DTC, wisdom of management technology*. http://dtcn-wisdom.jp/

Gilb, T. (2011). Estimation: A paradigm shift toward dynamic design-to-cost and radical management. *ASQ, SQP, 13*, 25–38.

Gilb, T., & Maier, M. W. (2005). *Managing priorities: A key to systematic decision-making*. Proceedings of INCOSE 2005. Available at http://www.gilb.com/tiki-download_file.php?fileId=60

Hipple, J. (2012). *The ideal final result—Optimization: The enemy of innovation*. New York, NY: Springer.

Ikovenko, S., & Bradley, J. (2003). TRIZ as a lean thinking tool. 2004 ETRIA TRIZ Future Conference, Florence, Italy. *TRIZ Journal*. Retrieved 17 August, 2015, from http://www.triz-journal.com/triz-lean-thinkingtool/

Isaka, Y. (2012). *Redesign by reducing the number of parts through TRIZ: Break away from existing designs and enhance competitiveness on cost*. Eighth TRIZ Symposium, Japan.

Li, T. (2010). Applying TRIZ and AHP to develop innovative design for automated assembly systems. *International Journal Advance Manufacturing Technology, 46*, 301–313.

Linstone, H. A. (2011). Three eras of technology foresight. *Technovation, 31*(2–3), 69–76.

Mann, D. L. (2002, May). Case study: The application of TRIZ to economy class aircraft cabin design. *TRIZ Journal*, Issues. Retrieved 17 August, 2015, from http://www.triz-journal.com/case-study-application-triz-economy-classaircraft-cabin-design/

Mann, D. (2002, June). Evolutionary-potential TM in technical and business systems [Issue]. *The TRIZ Journal*.

Mann, D. L., & Domb, E. (1999, May). 40 inventive (business) principles with examples. *TRIZ Journal*, Issues. Retrieved 17 August, 2015, from http://www.triz-journal.com/40-inventive-business-principles-examples/

Mann, D. L., & Domb, E. (2003). *Using TRIZ to overcome mass customization contradictions* (The customer centric enterprise, pp. 231–242). Berlin, Heidelberg: Springer.

Martin, A. (2010, September). TRIZ, theory of constraints and lean. *TRIZ Journal*, Issues. Retrieved 17 August, 2015, from http://www.triz.co.uk/files/triz_toc_and_lean.pdf

Montgomery, P., & Carlson, R. (2011). Control of total ownership costs of DoD acquisition development programs through integrated systems engineering processes and metrics. *Proceedings of Eighth Annual Acquisition Research Symposium, 8*(2), 63–77.

Rahim, Z. A., & Bakar, N. A. (2013). Implementation framework for design-to-cost using TRIZ: Product concept design in automotive stamping process. *American Journal of Economics, 3* (5C), 100–107.

Rahim, Z. A., & Nooh, A. B. (2014). Complexity planning for product design using TRIZ. *Advanced Materials Research, 903*, 396–401.

Rahim, Z. A., Sheng, I. L. S., & Nooh, A. B. (2015). TRIZ methodology for applied chemical engineering: A case study of new product development. *Chemical Engineering Research and Design, 103*, 11–24.

Sawaguchi, M. (2000, July). Effective approaches to solving technical problems by combining TRIZ with VE. *The TRIZ Journal*, Issue. Retrieved 17 August, 2015, from http://www.triz-journal.com/effective-approachessolving-technical-problems-combining-triz-ve/

Sheu, D. D., & Hou, C. T. (2011). *TRIZ-based problem solving for process-machine improvements: Slit-valve Innovative redesign* (pp. 1036–1044). Proceedings of the 41st International Conference on Computers and Industrial Engineering.

Stratton, R., & Mann, D. L. (2000, September). From trade-off to innovation—the underlying principles of TRIZ and TOC applied. System innovation and the underlying principles behind TRIZ and TOC applied to manufacturing. *The TRIZ Journal*, Issues. Retrieved 17 August, 2015, from http://scinnovation.cn/wpcontent/uploads/soft/100921/6-100921120609.pdf

Stratton, R., & Mann, D. L. (2003). Systematic innovation and the underlying principles behind TRIZ and TOC. *Journal of Materials Processing Technology, 139*, 120–126.

Tyson, K. W., Nelson, J. R., Om, N. I., & Palmer, P. R. (1989). *Acquiring major systems: Cost and schedule trends and acquisition initiative effectiveness (No. IDA-P-2201)*. Alexandria, VA: Institute For Defense Analyses.

Wichita, W. L. (1975). *Application of value engineering to design-to-cost contracts* (pp. 315–318). SAVE Conference Proceedings.

Wichita, L. W. (1980). *Design to cost contracts—Implementation of value engineering* (pp. 119–127). SAVE Conference Proceedings.

Williamson, B. (1994). *Design to cost lessons learned* (pp. 242–253). 1994 International Conference of the Society of American Value Engineers (SAVE): SAVE Annual Proceedings 1994.

Wu, T. D. (2004). *The study of problem solving by TRIZ and Taguchi methodology in automobile muffler designation*. Proceedings of the TRIZ Conference, Department of Industrial Management, Tung-Nan Institute of Technology, Taipei.

Zulhasni, A. R., & Nooh, A. B. (2015). Innovative cases of TRIZ application in the automotive industry. *Applied Mechanics and Materials, 735*, 326–330.

Zulhasni, A. R., Nooh, A. B., Sarimah, M., & Yeoh, T. S. (2015). TRIZ business improvement and innovation framework for malaysian small and medium enterprise. *Applied Mechanics and Materials, 735*, 349–353.

The Effectiveness of TRIZ Tools for Eco-Efficient Product Design

Issac Sing Sheng Lim

Abstract Eco-efficiency is a now necessity in view of rapidly depleting resources. Existing eco-efficiency tools provide standards and guidelines in developing product specifications that would have lesser environmental impact. However, there is a gap between setting those eco-efficient design targets and having the ability to achieve them. This research proposes the usage of tools based on the theory of inventive problem solving (TRIZ). The effectiveness of a variety of TRIZ tools in the eco-efficiency improvement of products will be studied. TRIZ tools such as the 39 Engineering Parameters, Contradiction Matrix, 40 Inventive Principles, Function Modelling, Trimming, Substance Field Modelling and 76 Standard Solutions are used in this research. A new tool termed the Eco Ideality Chart is also developed and used in this research. These tools were applied on actual products in collaboration with an electronics company. Three products which are a signal booster, streetlight and turnstile form the case studies. The design solutions developed by the company without the TRIZ tools were compared with the TRIZ-based solutions. It is observed that the TRIZ-based solutions did improve the eco-efficiency of the products. The product's functionality was increased with either the same or lesser resources. This research has shown the potential of TRIZ tools in the development of more eco-efficient products.

Keywords Eco-efficiency • TRIZ • Product design

1 Introduction

This research chooses to focus on the eco-efficiency of product design. In exact definition, eco-efficiency is achieved through the delivery of competitively priced goods and services that satisfy human needs and bring quality of life, while progressively reducing ecological impacts and resource intensity (Lim and Teoh 2010). The goods and services can be represented by products. Product functionality and ecological impact are determined by its design. The design stage of the

I.S.S. Lim (✉)
Monash University Malaysia, Sunway, Malaysia
e-mail: issaclss@gmail.com

L. Chechurin (ed.), *Research and Practice on the Theory of Inventive Problem Solving (TRIZ)*, DOI 10.1007/978-3-319-31782-3_3

product development determines the impact that the product will have throughout its life cycle to the environment.

The term eco-efficiency is originally coined by the World Business Council for Sustainable Development (WBCSD). Just like majority of the existing sustainability tools, eco-efficiency highlights the key areas where development can be done sustainably. The unique feature about the eco-efficiency approach is that it is developed and practised by companies within the WBCSD. Furthermore these companies are household brands that have a global reach in terms of their products and services. By being industry driven, the guidelines and standards proposed in eco-efficiency are more practical.

Overall, the philosophy of eco-efficiency is represented by seven elements. The first three eco-efficiency elements are on the reduction of material and energy and toxic dispersion. The following four eco-efficiency elements are on the increase of renewable resource usage, durability and service intensity.

All seven of the eco-efficiency elements are beneficial in identifying areas of improvements. Product specifications can be determined in the early stage of design to incorporate as many of the eco-efficiency elements. Currently more product design approaches are into setting more specifications that are oriented towards eco-efficiency (Russo et al. 2011a). These eco-efficiency elements are good guidelines on setting design targets. However, there is a lack of any guideline on how to achieve those targets.

An inventive problem-solving methodology known as the theory of inventive problem solving (TRIZ) is proposed in this research to assist the design of eco-efficient products. TRIZ is well established for overcoming contradictions and not compromising them. Likewise, there should not be a compromise on the ecological impact in pursuit of product functionality. TRIZ could be used to design more eco-efficient products. In fact, the first person to propose this is Genrich Altshuller who is the founder of TRIZ together with his research partner Michael Rubin back in 1991 (Shulyak and Rubin 1999).

The TRIZ tools used in this research work are those that are widely known and used. Some of the tools are meant to be used with each other in complement. Therefore, the tools are grouped into three sets. In the first set, the tools are the 39 Engineering Parameters, Contradiction Matrix and 40 Inventive Principles. For the next set, the tools are Function Modelling and Trimming. As for the final set, the Zone of Conflict, Substance Field Modelling and 76 Standard Solutions are the tools. All three sets of tools are used to develop systems with higher eco-efficiency.

The eco-efficiency improvements of systems are demonstrated through actual product case studies. A microelectronic company is engaged as the industrial research collaborator. This engineering-based company designs, develops and manufactures its own products. The role of the TRIZ tools in the improvement of the eco-efficiency in all three of the product case studies will be described.

2 Literature Review

There are pockets of research activities that are going on in relation to applying TRIZ in eco-efficient product design. The latest journals and conference papers related to this research topic have been reviewed. A summary of the 30 most relevant journals are summarized in the Table 1. The latest trends of the research related are identified and further explained in subsequent sections.

2.1 Tool Research Frequency

The TRIZ methodology consists of many tools. The frequency percentages of the various TRIZ tools used in eco-efficient research work are obtained by reviewing the latest related publications. Only journal and conference paper publications were considered as those are peer reviewed. As seen from Fig. 1, it is observed that not all of the tools are of similar popularity.

The tool which is most frequently researched on is the set of 40 Inventive Principles. At least 42 % of the total past research in eco-efficiency and TRIZ has used this tool. It does not come as a surprise as this tool is the most popular tool ever developed by the founder of TRIZ, Genrich Altshuller. There are multiple reasons for its popularity. The main reason is the tool's ease of use.

From the graph, it is observed that the more advance TRIZ tools are researched upon less. The advanced tools often require in depth knowledge and experience in using them. Therefore, not many researchers are equipped to apply them in research.

This research attempts to apply more TRIZ tools that previously conducted research works. All of the tools shown in the graph in Fig. 1 will be part for this research except for the 9 Windows, trends of system evolution and the algorithm of inventive problem solving (ARIZ).

2.2 Sets of Tools

Almost 90 % of the researches show that multiple TRIZ tools are used. This means that only TRIZ tools are meant to be used together. An average of 1.7 TRIZ tools have been used in each of the existing eco-efficiency related design research work. The most common tools used together are the 39 Engineering Parameters, Contradiction Matrix and the 40 Inventive Principles. Among the research work that utilizes multiple TRIZ tools, 30 % have chosen this set. For this research, more sets of tools will be used.

Table 1 Summary of researches related to sustainable approaches and TRIZ

Author	Sustainability approach				TRIZ tool										
	LCA	DFE	Eco-efficiency	ISO14001	9 Windows	Ideality	39 Engineering Parameters	Contradiction Matrix	40 Inventive Principles	Function model	Trimming	Substance field	76 Standard Solutions	Trends of system evolution	ARIZ
Low et al. (2000)							✓								
Liu and Chen (2001)			✓				✓		✓						
Strasser and Wimmer (2003)									✓						
Chang (2005)												✓	✓		
Kobayashi (2006)			✓				✓	✓	✓						
Turner (2009)			✓												
Lim and Teoh (2010)			✓						✓						
Chakroun et al. (2010)	✓											✓	✓	✓	
Trappey et al. (2011)	✓							✓				✓	✓		
Russo et al. (2011a)			✓												
Yang and Chen (2012)						✓	✓	✓	✓					✓	
Russo et al. (2011b)						✓					✓			✓	
Russo et al. (2011a)	✓								✓					✓	
Liu et al. (2011)	✓	✓												✓	
Chulvi and Vidal (2011)														✓	
D'Anna and Cascini (2011)					✓		✓	✓	✓					✓	

Yang and Chen (2012)				✓										✓	
Chang and Yeh (2012)	✓						✓	✓	✓						
Ferrer et al. (2012)	✓						✓	✓	✓						
Ferrer et al. (2012)	✓						✓	✓	✓						
Negny et al. (2012)							✓	✓	✓						
Cao et al. (2012)										✓					
Uang and Liu (2013)			✓				✓								
Kallel et al. (2013)	✓		✓				✓	✓	✓						
Mansoor et al. (2013)			✓						✓						
Lepeshev et al. (2013)		✓							✓				✓		
Hui et al. (2014)	✓											✓	✓	✓	✓
Russo et al. (2014)		✓							✓				✓	✓	
Chen and Jiao (2014)							✓	✓	✓						
Chen and Chen (2014)															✓
Mu and Zhao (2014)										✓		✓			

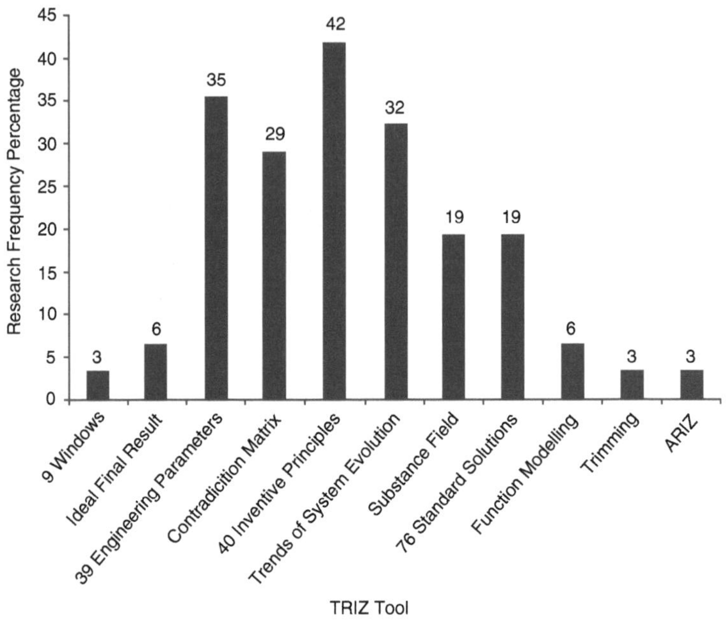

Fig. 1 Research frequency percentage of TRIZ tools graph

2.3 *New Tools*

A quarter of the latest research work on eco-efficiency and TRIZ introduces new tools. Most of these new tools are hybrids between eco-efficiency and TRIZ tools. This research also will use a new tool which is based on Ideal Final Result. The tool is being developed to qualitatively gauge the eco-efficiency of solutions.

2.4 *Case Study*

More than 80 % of the researches include case studies to enhance the findings of the research done. Case studies are effective in showing results that will support the claims of the research work. As seen in Fig. 2, product case studies are more popular choices compared to process or business model case studies. All three types of case studies can be eco-efficiently improved. In this research, the focus is on product case studies.

The case studies involved in this research work is in collaboration with a microelectronic company which designs and develops its own products. Therefore the case studies are based on actual products that are being developed by the company.

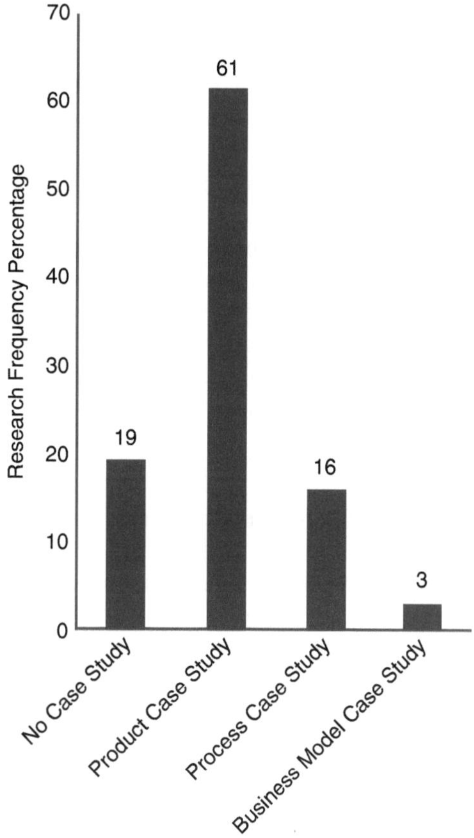

Fig. 2 Research frequency percentage of case study types graph

3 Methodology

In this research, both existing and newly developed TRIZ tools are used. The existing TRIZ tools are selected to eco-efficiently solve the design problems in the case studies. The eco-efficiency of the developed solutions is then gauged with a new iteration of a TRIZ tool. This tool is called Eco Ideality Chart.

The case studies selected for this research is in collaboration with a company which designs and manufactures its own electronic products. Hence, the company has full control on the product development process. The case studies chosen are related to actual products which have design problems. This research is conducted to observe if TRIZ tools can be used to develop solutions that deliver better product functionality and at a lower resource consumption.

3.1 *TRIZ Tool Selection*

From the literature review, it is observed that TRIZ tools have often been used without any specific sequence. Different TRIZ tools are effective in the different areas of problem solving. This research proposes the categorization of these areas into problem identification, problem exploration, solution generation and finally solution selection.

Each area of problem solving if done right will contribute towards a higher eco-efficiency for the final solution. For this research, the problem-solving steps in the case studies can be categorized into problem identification, problem exploration and solution generation. Each step uses different sets of TRIZ tools as summarized in Table 2. The role of each of the tools in increasing the eco-efficiency of a system is also described further.

3.1.1 Problem Identification

Determining the right problem to solve in the beginning of product design is important. The 39 Engineering Parameters, Function Analysis and Zone of Conflict assist in determining the actual root cause of the problem that needs solving.

39 Engineering Parameters The Engineering Parameters assist the establishment of a technical contradiction statement. It focuses the product developer to consider what parameter will worsen when another parameter is improved. Instead of optimizing to a compromise between the parameters, the contradiction needs to be removed.

Often times, when a product is designed to be eco-efficient, its functionality is compromised. Likewise when the functionality is increased, the eco-efficiency is compromised.

Function Type The overall functionality of a product relies on the sufficient functional interaction between the components. Components that are performing functions excessively are wasting resources and therefore should be reduced. On the other hand, functions that are insufficient or harmful will impede product functionality.

Table 2 Grouping of TRIZ tools for the case studies

Case study	Problem identification	Problem exploration	Solution generation	Solution selection
Signal booster	39 Engineering Parameters	Contradiction Matrix	40 Inventive Principles	Eco Ideality Chart
Streetlight	Function Analysis	Function Modelling	Trimming	Eco Ideality Chart
Turnstile	Zone of Conflict	Substance Field Modelling	76 Standard Solutions	Eco Ideality Chart

Sufficient functionality translates into efficient resource usage. Eco-efficiency of an overall system depends on the eco-efficiency improvements at the subsystem level.

Zone of Conflict The Zone of Conflict is the location where there is both the useful and also the harmful operating zone (Yeoh et al. 2009). It is the location in which the problem should be solved. Solving problems in just the useful or harmful zone might not be sufficient.

It is very often that the right solution could not be found because the problem was not the right one to begin with. This leads to the search for solutions for the wrong area which leads to more wasteful iterative steps.

3.1.2 Problem Exploration

Once the right problem has been identified, further exploration should be done on the problem. It is critical not to rush into developing solutions without a deeper understanding of the problem. The Contradiction Matrix, Function Modelling and Substance Field Modelling are able to provide different perspectives of the problem in the supersystem, system and subsystem level.

Contradiction Matrix The Contradiction Matrix recommends the most suitable Inventive Principles to each pair of contradicting Engineering Parameters. It is possible to consider multiple contradictions related to a single problem.

A contradiction might occur before or after the product is designed to be eco-efficient. When a system is designed to be more eco-efficient, there bound to be multiple contradictions. These eco-efficiency contradictions can be correlated to the Contradiction Matrix.

Function Modelling The Function Model will map out the interaction of all the subsystems and supersystems. Together with the Function Analysis, a complete understanding of the actual functional interactions that occur can be achieved.

Often times a new subsystem is introduced into a system to perform or improve a function. It is possible for any of the existing subsystem or supersystem to be used instead. With increased functionality and lower resource consumption, the eco-efficiency of the system is improved.

Substance Field Modelling A Substance Field Model shows the specific interaction between two subsystems. The functional interaction can either be sufficient, insufficient or harmful. The function field is also determined to be either of mechanical, thermal, chemical, electrical, gravitational or magnetic.

The types of functional interaction will determine the right solution that will be implemented. It is desired for the Substance Field interactions to be sufficient to ensure that the product's functionality is maintained or even improved.

3.1.3 Solution Generation

Conventional problem-solving method solely involves trial and error within the domain of expertise of the problem solver. Besides tools to identify the right problem and to explore the problem at a deeper level, TRIZ also consists of tools that suggest inventive solutions. The tools used in this research are the 40 Inventive Principles, Trimming and the 76 Standard Solutions.

40 Inventive Principles All patents represent solutions that have been developed to solve problems. The Inventive Principles are the synthesis of patents. Therefore the principles are proven solutions.

The Inventive Principles used to develop eco-efficient solutions are then proven as well. This means that lesser trial and error will be done. Wastage of resources is then prevented.

Trimming The purpose of Trimming is either to remove or to replace a subsystem (Yeoh 2014). The replacement can either be an existing subsystem or supersystem. Besides that, the replacement can be the introduction of a new subsystem.

Trimming is done on subsystems that perform inefficient, excessive or harmful functions. This ensures that the functionality of the overall system is sufficient. The system is kept minimally simple yet functional with continuous Trimming.

76 Standard Solutions All of the Standard Solutions are categorized into five classes. The different classes of standard solutions are meant to solve different types of substance field interaction.

The standard solutions generally recommend minimal change to the system in solving a problem. By following the suggestion on the Standard Solutions, lesser resources are needed to achieve a product functionality.

3.1.4 Solution Selection

In this research, the effectiveness of each TRIZ tool in developing an eco-efficient product is gauged with actual product case studies. To evaluate the eco-efficiency of the solution, a graphical representation of TRIZ's concept termed the Ideal Final Result is developed as shown in Fig. 3. It is termed as the Eco Ideality Chart. Solutions that have higher functionality and lower resource consumption are desired.

The purpose of the chart is to provide a simple qualitative eco-efficiency comparison of the previous and newly developed product. Ideal Final Result is about developing solutions with higher functionality at the least amount of cost or harm. The terms cost and harm are represented as resource consumption in the Eco Ideality Chart. There is consumption with usage and also harmful damage of resources. This chart provides a simple overview comparison of the solutions generated during product development.

Fig. 3 Eco Ideality Chart

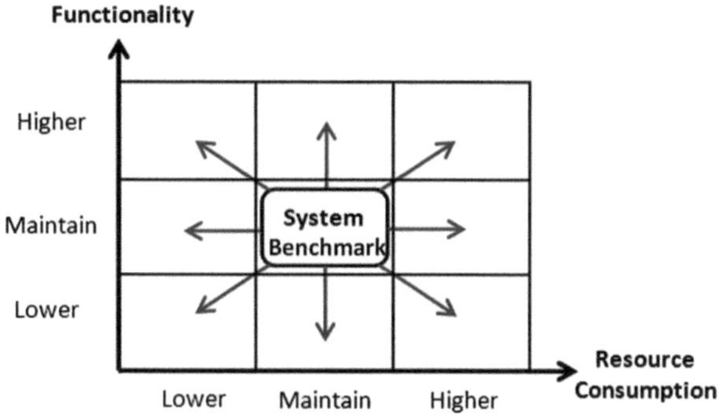

4 Results

The effectiveness of TRIZ in developing eco-efficient design is tested through three product case studies. These case studies were conducted in collaboration with a microelectronics company that develops its own products.

All of the case studies did not have any design specifications related to eco-efficiency. They begin with a design problem that needed to be solved. For each of these product design case study, different sets of TRIZ tools are used to solve the problems.

In each of the case study, the original solutions developed by the product development team are first presented. After that the step-by-step explanation of the TRIZ tools used is provided. Next, the TRIZ-based solutions are also presented. Comparisons between the two types of solutions in terms of eco-efficiency will also be discussed.

4.1 Signal Transmitter

The first product case study discussed here is a signal transmitter. Analogue signals are converted into digital signals and then transmitted using a cable. This enables integration of analogue and digital systems. Similar products are already available in the market by other competing companies.

To be different, the company in collaboration with this research developed a much smaller signal transmitter. It is to be less than a quarter of the existing product's size. Not only less material is needed, its size would also mean that it can be placed in strategic places.

Within the transmitter, heat is being transferred by the electronic components. The existing transmitter in the market has an inbuilt fan to dissipate the heat. However as the newly developed product is smaller, a fan could not be placed inside the product to cool it down. The increased heat affected the performance of

Table 3 Contradiction Matrix for the signal booster case study

Improving parameter	Worsening parameter		
	27. Reliability	37. Difficulties of detecting and measuring	39. Productivity
8. Volume of stationary object	2, 16, 35	2, 17, 26	2, 10, 35, 37

the product in terms of transmittance quality and speed. Therefore, heat dissipation has to be improved without a fan.

First of all a contradiction statement is formed using the 39 Engineering Parameters. As shown in Table 3, the main parameter that has been improved is the volume of the transmitter. By improving that parameter, a few other parameters such as reliability, measurement ability and productivity have worsened.

Based on these parameters, the Contradiction Matrix is referred to. A total of seven Inventive Principles were recommended by the matrix. From there, four design solutions were developed.

With Inventive Principle 10 of Prior Action, a thermal tape is placed over the electronic components that emit the most amount of heat. The thermal tape functions as a conductor that quickly transfers heat away.

Next, Inventive Principle 16 of Partial or Excessive Action was applied on the selection of the microprocessors. The microprocessors were programmed not to work at the default maximum capacity at all times. With this, less heat is generated. With lesser heat generation, the performance of the device would not be adversely affected.

Finally, two solutions were developed using Inventive Principle 17 of New Dimension. The circuit board is now designed to be multilayered. Each layer has the electronic components facing outwards towards the casing as shown in Fig. 4. Therefore more heat emitted by the electronic components is transferred to the casing which acts as a heat sink.

A total of four solutions were developed with the aid of the 39 Engineering Parameters, Contradiction Matrix and 40 Inventive Principles. All of the solutions were implemented in the final product.

4.2 *Streetlight*

The second product case study covered is the streetlight. Conventional streetlights use high-pressured sodium lamps. The newly developed streetlight uses light-emitting diodes (LEDs) as they are more energy efficient. Though the streetlight is more energy efficient, the lifespan of the LEDs were shortened due to the high operating temperature.

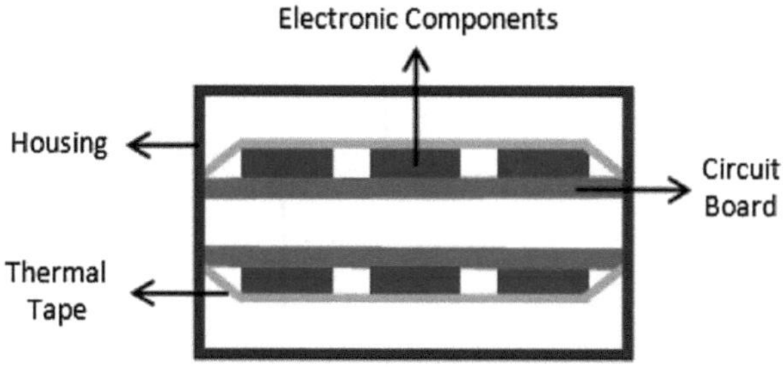

Fig. 4 Cross-sectional area of the signal booster

For cooling, the current design is using aluminium cooling fins. The cooling fins are attached to the LED circuit board to direct heat away. Larger fins correlate to more heat dissipation. However with larger fins, the total weight exceeds the allowable load that can be held by the streetlight pole.

A Function Model of the streetlight is developed as shown in Fig. 5. The functional interaction between the main components was also identified. While the housing of the LEDs could sufficiently hold the circuit board and cooling fins in place, the pole insufficiently holds the extra weight of the total housing weight that includes the cooling fins. Even though more cooling fins are used, the LED circuit board is still insufficiently cooled. Thus reducing the lifespan of the LED.

Since the cooling fins insufficiently performs the cooling function and causes excessive load, it is to be trimmed. Instead of optimizing the cooling fins, a new lightweight component that could cool down the circuit board should be searched. A market search has led to the identification of a newly developed material that is an excellent heat conductor and also lightweight.

The material is graphite and it is being marketed in the form of sheets. A proof of concept was established to indicate the suitability of the graphite sheet for application in this product. The graphite sheet did reduce the temperature of the LEDs. The overall weight load for the streetlight pole is also met. Therefore, all of the functions between the components are sufficient as shown in Fig. 6.

4.3 *Turnstile*

In the final case study, a turnstile will be discussed. Turnstiles function as gates to control large human traffic flow at entrances to both open and close spaces. Instead of being manually operated, turnstiles are now fully automated. Fully automated turnstiles are a common feature especially in modern business buildings. The company in collaboration in this research has also developed a new type of turnstile. In contrast to the standard turnstiles which are available in the market, it is designed to be compact and lightweight.

During the pilot testing of this new turnstile, a design problem was detected. Inside the turnstile casing, there are two circuit boards that are facing each other. The top circuit board contains infrared transmitters, whereas the infrared receivers

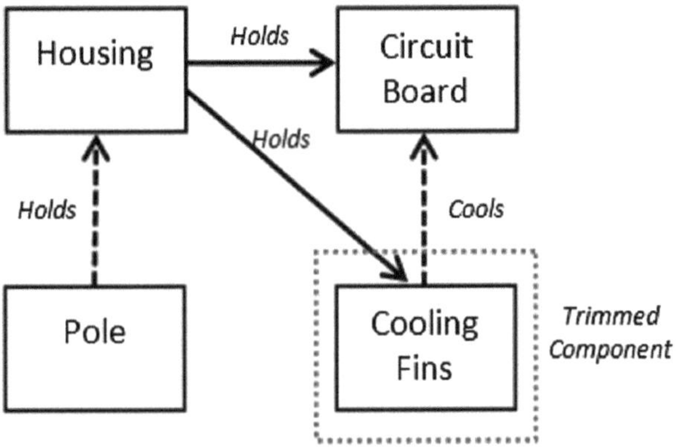

Fig. 5 Function model of the original streetlight

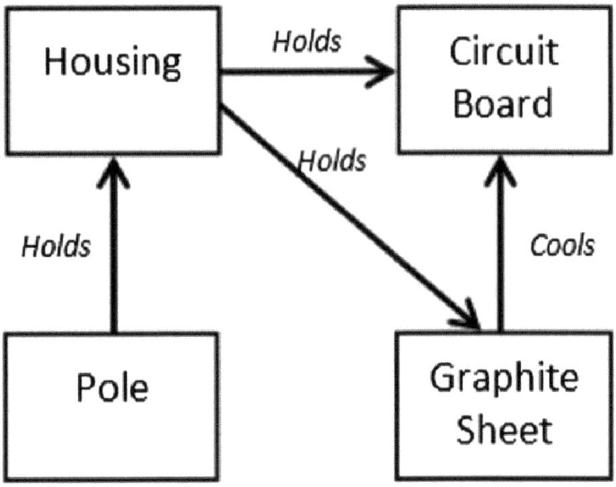

Fig. 6 Function model of the new streetlight

are at the bottom circuit board. Both types of transmitters should not be in contact with each other and should always be parallel to enable the transmission of signals. The signals are used to control the opening and closing of the turnstile flaps.

However, the top circuit board is found to be tilted. The tilt is caused directly by the rotating shaft of which the top circuit board is attached to. This alters the position of the infrared transmitters and receivers. Thus, affecting the product's ability to open and close the turnstile flaps.

It is not possible to place the circuit boards further away from each other as the height clearance inside the casing is very limited. Also, no major change of components is allowed. An immediate solution is needed to solve flap movement problem.

Initially the focus of the problem was on the tight area between the lower casing and the shaft. Due to the casing being very thin, it could not hold the shaft in position. There are conceptual solutions developed by the engineering team without using TRIZ. It was proposed that the lower casing be made thicker so it could grip

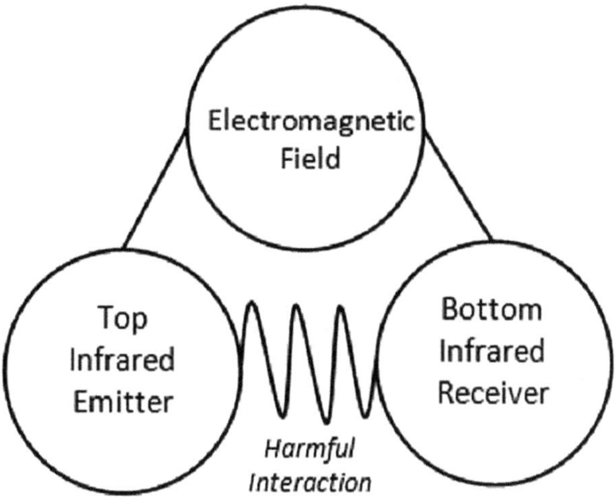

Fig. 7 Substance-field model of the turnstile's zone of conflict

and hold the shaft in position. Alternatively, a ring was proposed to be soldered to the bottom of the casing where the shaft enters. This could not be applied as there should have enough clearance to allow the shaft to rotate within the ring's centre. Both solutions could not solve the problem and they incur extra costs.

With TRIZ, the Zone of Conflict was found to be the space between the two circuit boards. The space is useful for transmission of signals, but it causes harm as it allows the infrared transmitters and receivers to have contact. Therefore a solution was focused on this area instead of the shaft.

In Fig. 7, Substance Field Analysis shows a harmful interaction between the infrared emitter and receiver. Then, the 76 Standard Solutions is referred. As the Substance Field is of harmful interaction, Class 1.2 was referred. Immediately the Class of 1.2.1 tells that an extra substance is needed to block the harm between the original two substances.

A cost-effective and instantly applicable solution was developed in the form of a silicone rubber piece. The silicone rubber is placed in between the top and bottom infrared circuits as shown in Fig. 8. This piece acts as a stopper that prevents the top circuit board from tilting downwards. Instead of redesigning the shaft and casing design, this stopper can be fixed instantly at almost no cost as it is very cheap.

5 Discussion

There were two types of solutions being presented in each of the case study. The first being the original solution and the second is TRIZ based. Different observations are made for both types of solutions. These are three main observations of the original solutions by the product developers in the collaborating company.

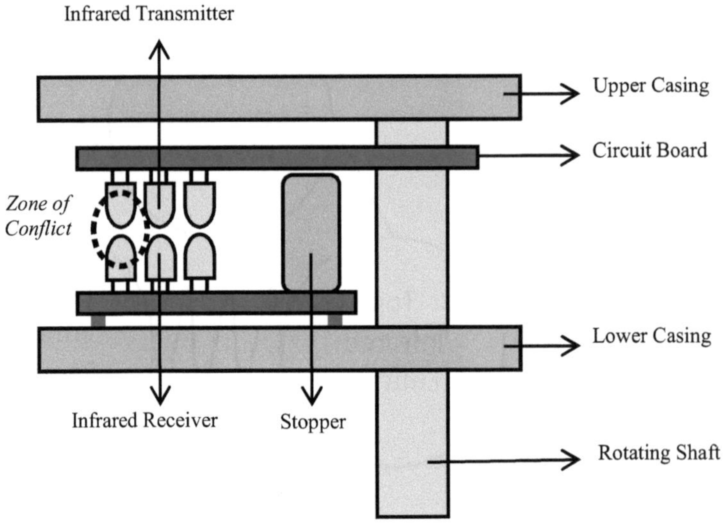

Fig. 8 Cross-sectional view of a part of the turnstile

Followed on by each of this observation, the TRIZ-based solutions are described to highlight the differences between them.

Firstly, there are solutions that can solve the problem but with the usage of more resources. This is shown in the case study on the signal booster. The heating problem is originally solved using a fan which is an extra component to the product, and it consumes more power to operate. With the recommendation of the Contradiction Matrix, a few Inventive Principles were used to develop a few solutions. The solutions developed increased the cooling functionality without a fan.

For the second observation, the solutions proposed could not improve the existing system. The solution performs the same level of function but with more resources. This is depicted in the usage of more cooling fins for the streetlight. There are limitations to the amount of heat that can be dissipated efficiently by the cooling fins. Hence, more resources are consumed to achieve the same level of functionality. Instead of focusing on optimizing the aluminium cooling fins, the Function Model of the streetlight has led to the trimming of the fins. Graphite sheets are used to dissipate heat instead. Lesser resources are used as the graphite's heat conductivity to weight ratio is higher than that of the aluminium cooling fins.

Finally, the third observation made is that it is possible to develop solutions that are worse than before. A bad solution aggravates the problem further and consumes even more resources. In preventing the rotating shaft that holds the flap doors of the turnstile from tilting, the original solution was to make the cover thicker and introduce a ring. However, the shaft does not have enough clearance to rotate. Hence, a new problem occurs and more resources are used. With the suggestions from the 76 Standard Solutions, a simple rubber stopper was placed at the right Zone of Conflict. The turnstile's functionality improved significantly at almost no cost.

Table 4 Function and resource consumption comparison

Case study	Function		Resource consumption	
	Without TRIZ	With TRIZ	Without TRIZ	With TRIZ
Signal booster	Boosts and sends signal (function improved)	Boosts and sends signal (function improved)	Cooling fan (resource consumption increased)	Fanless (resource consumption reduced)
Streetlight	Cools circuit board (function maintained)	Cools circuit board (function maintained)	More cooling fins (resource consumption increased)	Graphene sheets (resource consumption reduced)
Turnstile	Opens and closes door incompletely (function worsened)	Opens and closes door completely (function improved)	Metal ring and clamp consumption increased)	Silicone stopper (resource consumption reduced)

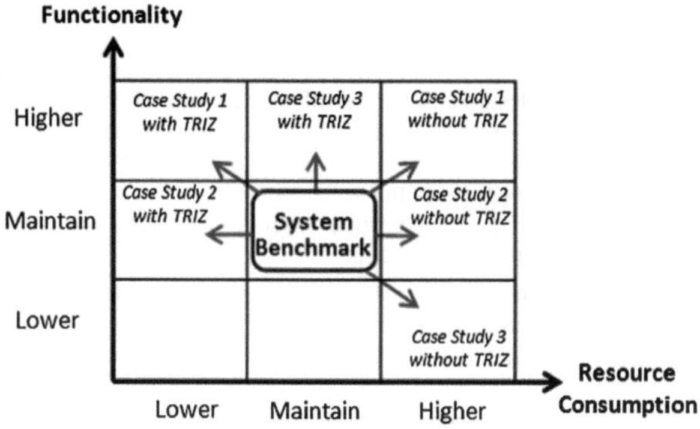

Fig. 9 Eco Ideality Chart for the case study solutions

Overall, the TRIZ-based solutions are more resource efficient relative to the previous system. More importantly, the TRIZ-based solutions improve product functionality by solving contradictions instead of optimizing. Hence, the new systems are more eco-efficient as tabulated in Table 4 and as shown through the Eco Ideality Chart in Fig. 9. Without TRIZ, there will be higher consumption of resources. Even with more resources, product functionality might not necessarily increase and might be worsened. Therefore through the case studies, the different levels of eco-efficiency of non-TRIZ and TRIZ-based solutions can be compared.

6 Conclusion

The design of a product will determine its functionality and also environmental impact through the resources consumed. Eco-efficiency is the ratio between the product's functionality and impact to the environment. In this research, three actual industrial case studies have been conducted on a signal booster, streetlight and turnstile. Solutions using and not using TRIZ were described in detail. TRIZ not only solve design problems effectively but also it solves them more eco-efficiently.

As shown in the Eco Ideality Chart, the solutions developed based on TRIZ are more eco-efficient. Though the eco-efficient products have not reached the Ideal Final Result, they are a step closer towards being resource independent. This research work highlights that the various classical TRIZ tools are effective in solving design problems and also in improving product eco-efficiency.

References

Cao, G. Z., Guo, H. X., Zhang, C. Y., & Li, Q. H. (2012, March). Research on sustainable function innovation based on TRIZ. *Applied Mechanics and Materials, 148*, 1429–1433.

Chakroun, M., Gogu, G., Pradel, M., Thirion, F., & Lacour, S. (2010, April). *Eco-design in the field of spreading technologies*. Green Technologies Conference, 2010 I.E. (pp. 1–6). IEEE.

Chang, H. T. (2005, December). *The study of integrating Su-field analysis modeling with eco-innovative concept for product design*. 4th International Symposium on Environmentally Conscious Design and Inverse Manufacturing, 2005 (Eco Design 2005) (pp. 663–670). IEEE.

Chang, M. T., & Yeh, C. H. (2012). An innovative manufacturing process for bamboo injection molding by using TRIZ and Taguchi method. *Journal of the Chinese Institute of Industrial Engineers, 29*(7), 454–465.

Chen, W. C., & Chen, J. L. (2014). Eco-innovation by integrating biomimetic design and ARIZ. *Procedia CIRP, 15*, 401–406.

Chen, J. L., & Jiao, W. S. (2014). TRIZ innovative design method for eco-leasing type product service systems. *Procedia CIRP, 15*, 391–394.

Chulvi, V., & Vidal, R. (2011). Usefulness of evolution lines in eco-design. *Procedia Engineering, 9*, 135–144.

D'Anna, W., & Cascini, G. (2011). Supporting sustainable innovation through TRIZ system thinking. *Procedia Engineering, 9*, 145–156.

Ferrer, J. B., Negny, S., Robles, G. C., & Le Lann, J. M. (2012). Eco-innovative design method for process engineering. *Computers and Chemical Engineering, 45*, 137–151.

Hui, Y., Guo, J., & Fan, D. L. (2014). Integrated TRIZ and MEMS in eco-innovative design. *Applied Mechanics and Materials, 441*.

Kallel, W. S., Ledoux, Y., & Nadeau, J. P. (2013, May). *Eco-innovative method to improve the distribution phase of product*. Advanced Logistics and Transport (ICALT), 2013 International Conference on (pp. 529–534). IEEE.

Kobayashi, H. (2006). A systematic approach to eco-innovative product design based on life cycle planning. *Advanced Engineering Informatics, 20*(2), 113–125.

Lepeshev, A. A., Podlesnyi, S. A., Pogrebnaya, T. V., Kozlov, A. V., & Sidorkina, O. V. (2013, September). *TRIZ-based engineering education for sustainable development*. Interactive Collaborative Learning (ICL), 2013 International Conference on (pp. 489–493). IEEE.

Lim, I. S. S., & Teoh, K. S. (2010). Eco-efficient product design using theory of inventive problem solving (TRIZ) principles. *American Journal of Applied Sciences, 7*(6), 852.

Liu, C. C., & Chen, J. L. (2001, December). *Development of product green innovation design method*. Ecodesign (p. 168). IEEE.

Liu, Z. F., Gao, Y., Zhang, L., & Hu, D. (2011, September). Green innovation design based on TRIZ theory. *Advanced Materials Research, 308*, 373–379.

Low, M. K., Lamvik, T., Walsh, K., & Myklebust, O. (2000). *Product to service eco-innovation: the TRIZ model of creativity explored*. Electronics and the Environment, 2000. ISEE 2000. Proceedings of the 2000 I.E. International Symposium on (pp. 209–214). IEEE.

Mansoor, M., Mariun, N., Ismail, N., & Wahab, N. I. A. (2013). A guidance chart for most probable solution directions in sustainable energy developments. *Renewable and Sustainable Energy Reviews, 24*, 306–313.

Mu, R., & Zhao, N. (2014, May). Design for disassembly based on substance-field analysis. *Applied Mechanics and Materials, 527*, 105–110.

Negny, S., Belaud, J. P., Robles, G. C., Reyes, E. R., & Ferrer, J. B. (2012). Toward an eco-innovative method based on a better use of resources: application to chemical process preliminary design. *Journal of Cleaner Production, 32*, 101–113.

Russo, D., Bersano, G., Birolini, V., & Uhl, R. (2011a). European testing of the efficiency of TRIZ in eco-innovation projects for manufacturing SMEs. *Procedia Engineering, 9*, 157–171.

Russo, D., Regazzoni, D., & Montecchi, T. (2011b). Eco-design with TRIZ laws of evolution. *Procedia Engineering, 9*, 311–322.

Russo, D., Rizzi, C., & Montelisciani, G. (2014). Inventive guidelines for a TRIZ-based eco-design matrix. *Journal of Cleaner Production, 76*, 95–105.

Shulyak, L., & Rubin, M. (1999). What will happen after the final victory? *Journal for the Altshuller Institute for TRIZ Studies*, Volume 1, Autumn 1999, 1–9.

Strasser, C., & Wimmer, W. (2003, December). *Supporting customer driven eco-solutions-implementing ecodesign in the daily work of product developers*. Environmentally Conscious Design and Inverse Manufacturing, 2003. EcoDesign'03. 2003 3rd International Symposium on (pp. 757–762). IEEE.

Trappey, A. J., Ou, J. J., Lin, G. Y., & Chen, M. Y. (2011). An eco-and inno-product design system applying integrated and intelligent QFDE and TRIZ methodology. *Journal of Systems Science and Systems Engineering, 20*(4), 443–459.

Turner, S. (2009). ASIT—A problem solving strategy for education and eco-friendly sustainable design. *International Journal of Technology and Design Education, 19*(2), 221–235.

Uang, S. T., & Liu, C. L. (2013). The development of an innovative design process for eco-efficient green products. In M. Kurosu (Ed.), *Human–computer interaction. Users and contexts of use* (pp. 475–483). Berlin, Heidelberg: Springer.

Yang, C. J., & Chen, J. L. (2012). Forecasting the design of eco-products by integrating TRIZ evolution patterns with CBR and simple LCA methods. *Expert Systems with Applications, 39*(3), 2884–2892.

Yeoh, T. S. (2014). *Systematic innovation in business and management* (p. 105). Selangor, Malaysia: First Fruits Publishing.

Yeoh, T. S., Yeoh, T. J., & Song, C. L. (2009). *TRIZ-systematic innovation in manufacturing* (p. 70). Selangor, Malaysia: First Fruits Publishing.

Using Enhanced Nested Function Models for Strategic Product Development

Horst Th. Nähler and Barbara Gronauer

Abstract TRIZ provides excellent tools for designing customized problem-solving and product developing processes and algorithms. Strategic decisions can furthermore be strengthened by tools still in development and under research, as the Trends of Engineering Systems Evolution and their underlying mechanisms. One of the difficulties when using TRIZ tools is the large number of possibilities for using and combining them for the best effect. Especially during the analysis phase, a lot of information has to be gathered that leads to problem models and task definition for later problem solving. To expand the usage of analytical tools for strategic decisions for long-term disruptive innovations or short-term incremental innovations, this chapter proposes a scientific approach that is based on the Theory of Inventive Problem Solving and its findings. The described procedure helps companies to find the right long- or short-term decisions:

(a) They can check when their product might be eliminated from the market by following products and new settings and define indicators for long-term decisions, what the next products could be;
(b) Similar to the forecast procedure the incremental way of further product development can also be checked and with this overview the development steps for the R&D, production, and sales departments can be better planned.

The approach combines proven tools like Function Analysis with a new approach for modeling complex engineering systems (Nested Function Models), 9-Screen Models, S-Curve Analysis, and the Trends of Engineering System Evolution. This chapter examines the interactions between components on different system levels, the use of the model in conjunction with trimming, and the integration of the Multi-Screen Approach for developing a basis for strategic product development decisions. Furthermore, the possibilities of connecting Trends of Engineering Systems with this approach are explored. The suggested approach is

H.T. Nähler (✉)
c4pi – Center for Product-Innovation, Hünfeld, Germany
e-mail: naehler@c4pi.de

B. Gronauer
StrategieInnovation, Hünfeld, Germany
e-mail: bg@strategieinnovation.de

L. Chechurin (ed.), *Research and Practice on the Theory of Inventive Problem Solving (TRIZ)*, DOI 10.1007/978-3-319-31782-3_4

aimed at creating an extensive, multilevel product map that combines the benefits of several classical TRIZ tools. It also creates a base for strategic decisions linked to problem-solving opportunities on the operative level.

Keywords Strategic Planning • Foresight • Function Model • 9-Screen Model • Trends of Engineering System Evolution • Systematic Innovation • TRIZ

1 Introduction

One of the aspects that draw people to the Theory of Inventive Problem Solving is the analytical approach to inventive problems on one hand and on the other hand to get an idea about the life cycle of their product and the next big development steps.

Having a structured way of exploring an engineering system, revealing its shortcomings, and extracting current limits that prevent the system to develop further are the main benefits of TRIZ.

The experience from several industrial projects and feedback from practitioners gathered on conferences brought up the need to be able to assess a complex engineering system (Cavallucci et al. 2014; Nähler et al. 2012) with Function Analysis without the need to focus on a certain system component or to model the system on a too general high level. Instead of creating a number of single Function Models for each desired abstraction level, an expansion of the Function Model is proposed, in which function models are nested inside each other to create a complete model of an engineering system from a high abstraction level down to each part on the lowest abstraction level.

The proposed approach to create such "Nested Function Models" for complex systems is also capable to serve as a strategic tool, as it creates a complete product map for any engineering system. The combination with other strategic tools like S-Curve Analysis and the Trends of Engineering System Evolution (TESE) seems useful. The 9-Screen model serves as a starting point and a connecting, underlying structure for the creation of a product map that integrates nested function models, S-Curve Analysis, and TESE.

This chapter proposes the combination and integration of well-known and proven TRIZ tools without violating the rules and directions for their application. Thus, users who are familiar with the body of knowledge of TRIZ can benefit directly from the approach described in this article regarding the following aspects:

1. Greater confidence for planning future product generations
2. Gain of time for development decisions
3. Assessing an engineering system in its entirety
4. Focused development of customer value with respect to the S-Curve

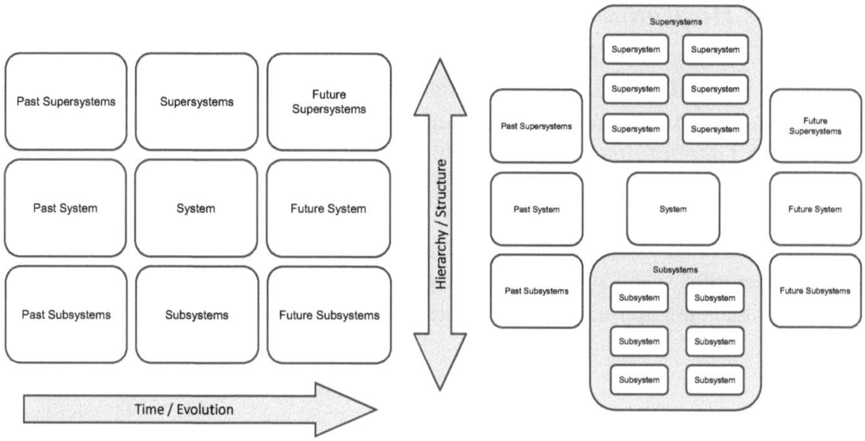

Fig. 1 9-screen model

2 The 9-Screen Model as a Generic Product Map

As a universal scheme to structure the composition of an engineering system in the context of its past and future, this classical TRIZ Tool is easily adopted by engineers. Its straightforward approach makes it extremely useful to accomplish numerous tasks: from creating future product concepts and usage scenarios to identifying new business opportunities or making strategic decisions on an enterprise level (Souchkov 2014; Altschuller 1998; Altshuller 2000) (Fig. 1).

2.1 The Vertical Axis of the 9-Screen Model

The vertical levels of the 9-Screen Model represent the structure of an engineering system with its components (subsystems) and defines its boundaries to surrounding or neighboring systems (supersystems).

The middle screen—the system that is to be assessed—has to be defined by the user. This demarcation of the system and surrounding systems also sets the perspective about the components that make up the system.

It is obvious that the 9-Screen Model can be expanded in the vertical to represent hierarchical structures, similar to assembly trees in CAD models. Subsystems might also contain sub-subsystems and so on. As a necessary step within Function Analysis, the Component Analysis results in similar schemes as the 9-Screen Model, making it easy to combine the two tools.

2.2 *The Horizontal Axis of the 9-Screen Model*

While the vertical axis shows the arrangement of the system, the horizontal axis represents the timeline related to either the system's history and future or the "Operating Time" during which the problem happens and the time before and after.

Using the 9-Screen Model in strategic context, we can assess every system, big or small, and research and describe its history as well as the evolution of relevant supersystems and system components. So no matter how deep we go down into a product hierarchy, each subassembly and part has its own history and future. This aspect enriches the approach of the Nested Function Model described below.

3 Nested Function Models

Usually the function analysis process begins with a component analysis, where the system level and thus the abstraction level are defined. The criteria to make a decision on the choice of an appropriate abstraction level depends on the projects' goals and the intended outcome. Problem Solving on the level of a subassembly or a single part of the system might result in a low level of abstraction and therefore a refocus and redefinition of the term "system." Usually, it is not recommended and not practical to model a complete System using all of its single parts as components, as the resulting function model becomes increasingly cluttered and confusing (Training Material MATRIZ 2006, 2010) (Fig. 2).

To overcome this drawback, it is suggested to use the 9-Screen philosophy and look at each component as a "box" that again contains components, which might

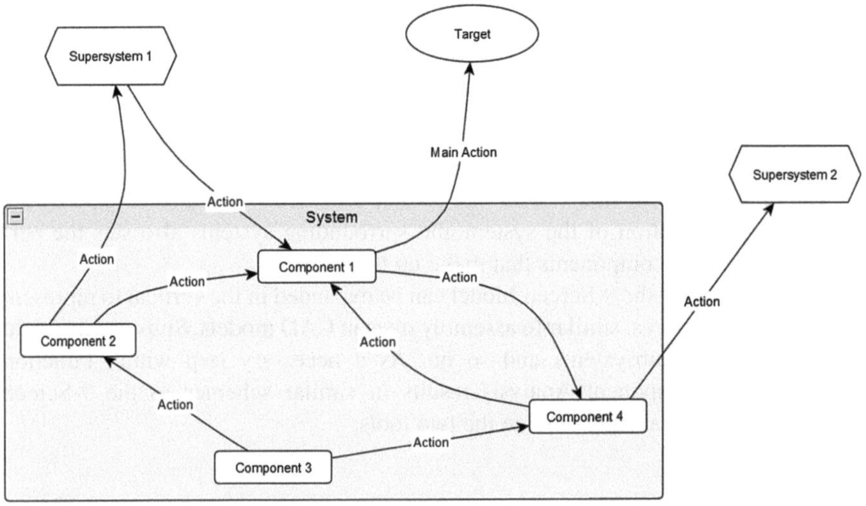

Fig. 2 Generic function model structure

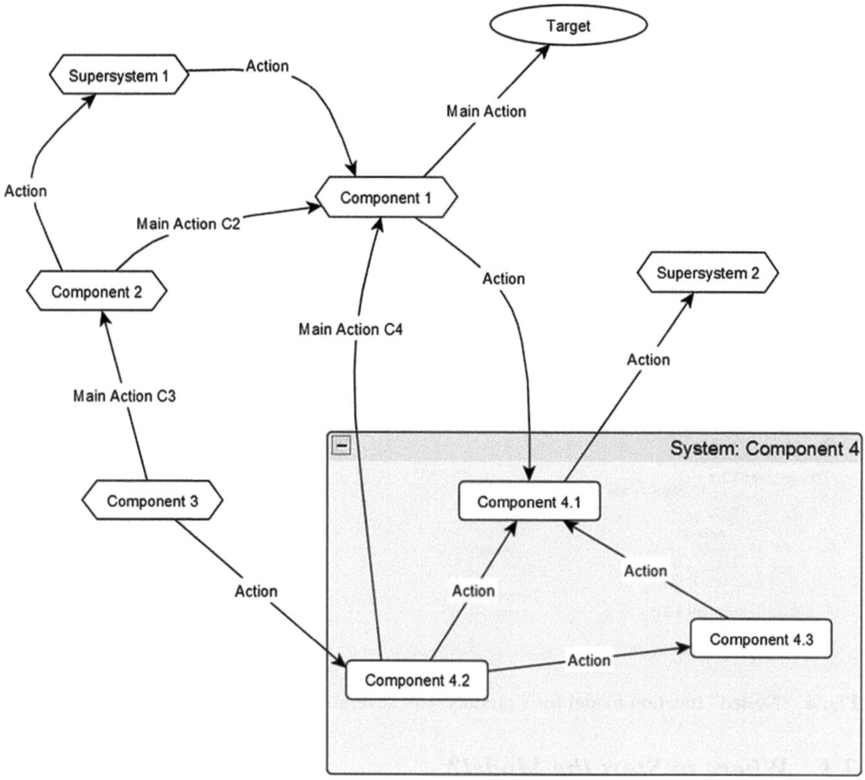

Fig. 3 The nature of refocusing according to 9-screen model: When component 4 is considered the system, each of the other components becomes supersystems and parts of component 4 become components of the system

again consist of several subcomponents. We are then able to build function models for each component and nesting them to generate a hierarchical structure (Fig. 3).

Following this process we can generate a complete product map without cluttering it, as we can chose which component or system to assess. We can also track down the components on a low abstraction level which are involved in the main function on a high product level. This enables us to recognize if main function (s) are affected by changes on (sub-. . .) subsystems.

Furthermore, it is possible to delegate the work of generating function models of subassemblies to different teams and then "plug" the single models inside each other to complete the product map (Fig. 4).

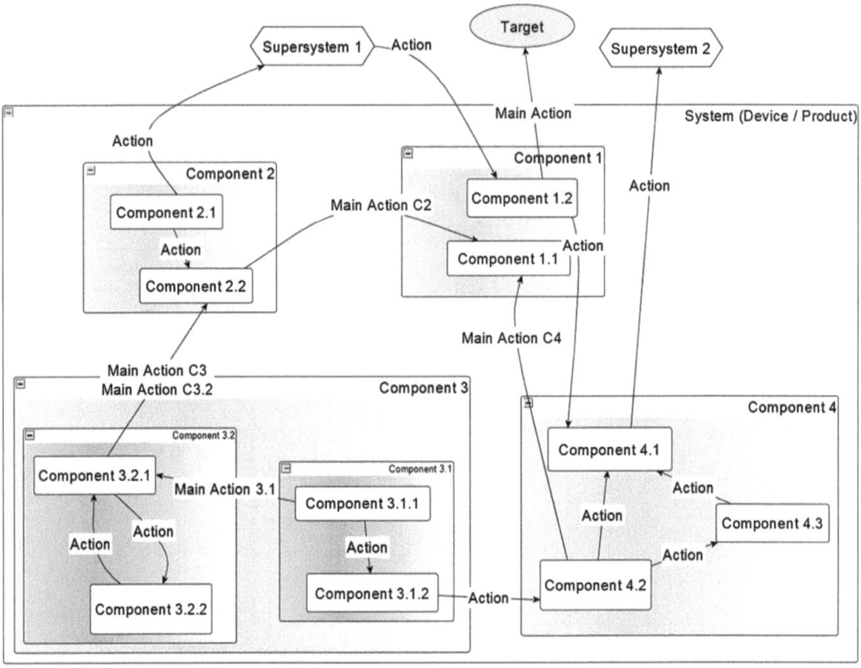

Fig. 4 "Nested" function model for a product with several subassemblies

3.1 Where to Start the Model?

In Function Analysis, the first step necessary is to define the system, its components (subsystems), and supersystems. Of course, we can start the Nested Function Model on each hierarchical level, but it is recommended to start from the perspective of the Product, e.g., the whole product that is sold to customers. When dealing with systems on a bigger scale (e.g., a power plant) the question generally can remain the same: Which system generates an output that creates a value for the customer?

Also, the modeling scope varies with the purpose of the analysis. For incremental "optimizations" or small-scale innovations, a thorough analysis of the subsystems of a product is important while high-level, disruptive innovations need an extensive consideration of the supersystems. Depending on the task, the focus is set on either subsystems or supersystems.

3.2 How Many Hierarchical Levels Should Be Considered?

If incremental development of a product is desired, in most cases the analysis of the subsystems is a top priority. But if we need to assess complex systems, e.g., a whole car, the product considered might be made up of a huge number of single components/parts. Of course, we are not limited with regard to the "last" or deepest system level to model. When modeling an electronic device, we certainly can go down to the components of the capacitor or resistor or even down to the chemical components of the dielectric. Practical considerations should guide the user here. Usually, stopping at single parts that are bought from a supplier as the last level of the Nested Function Model is a good practice. Also, subassemblies that are acquired from system suppliers might be a good indicator not to go into further detail (except a combined effort to develop the subsystems in cooperation with the supplier is desired).

If the aim is to achieve disruptive solutions, it is necessary to do a function model with more supersystems taken into consideration. It is highly recommended to assess the whole process in which the product is embedded with all systems involved. For disruptive concepts, these supersystems are of great value while sub-(sub-)systems might quickly be rendered unnecessary when a completely new technology is used in a disruptive solution.

3.3 *Thoughts on Main Functions*

During this component Analysis, it is necessary to formulate the Main Function. Following the definition, the Main Function acts on the Target and represents the "reason" why the system has been created or what the system actually does (Hentschel et al. 2010). Following the rules of TRIZ Function Analysis (Training Material MATRIZ 2006, 2010), the target is always to be found in the supersystem or the "neighborhood" of the system, so when we use Nested Function Models, the main function of each subassembly acts on a component of a higher system hierarchy.

Throughout the Nested Function Model, we need to define the Main function of each sub- or sub-sub-assembly down to each single part.

3.4 *Defining the Time frame for the Nested Function Model*

According to TRIZ, defining the Operating Time is good practice (Altshuller 2000). As a starting point it is suggested to look at the time when the system carries out or delivers its Main Function (Koltze and Souchkov 2011). As the Nested Function Model mainly deals with strategic evaluation, (exclusively) modeling problem situations is not a primary focus.

The point in time or the time frame chosen should be consistent throughout the levels of the Nested Function Model. One problem arises when different usage scenarios are possible and the product can be used in different situations. In these cases, it is useful to define those usage scenarios and build a model for each of them.

Many systems contain subsystems that only exist to act during specific situations or usage scenarios. In Function Models, we can easily identify these components, as they are only “receiving” actions during the “normal” use of the system and are therefore only Function Objects. Those cases might be problem situations (e.g., emergency siren: During the normal use of a machine, the siren does not “emit” any action and is therefore not function carrier, it is only an object) or a handling robot inside a processing machine that is active only during specific operations.

Of course, we can always ask the question: “What is the time when the main function of the (e.g., emergency- or handling-) subsystem is carried out?” and model these time frames. This would usually lead to an inconsistent definition of the time frame and it would clutter the Function Model because components on higher system hierarchy might behave differently in those situations as well.

4 Enhancing Nested Function Models with the Time Axis

As described in Sect. 2.2, looking into the history of each Subsystem enables us to assess each component with regard to possible future developments. If a Nested Function Model already exists, the definition of supersystems and subsystems is already at hand. We should then be able to research past versions of the “system” under evaluation, or if the system has been nonexistent in the past and has been added to increase functionality. Also, we can see which “past components” have been trimmed and to which components, e.g., on a higher hierarchical level, the main functions have been transferred to.

Additionally, we can even model past versions of each component (system or subsystem or sub-subsystem...) and therefore create a functional evolutionary map that gives us an overview of the structural evolution of the system (Fig. 5).

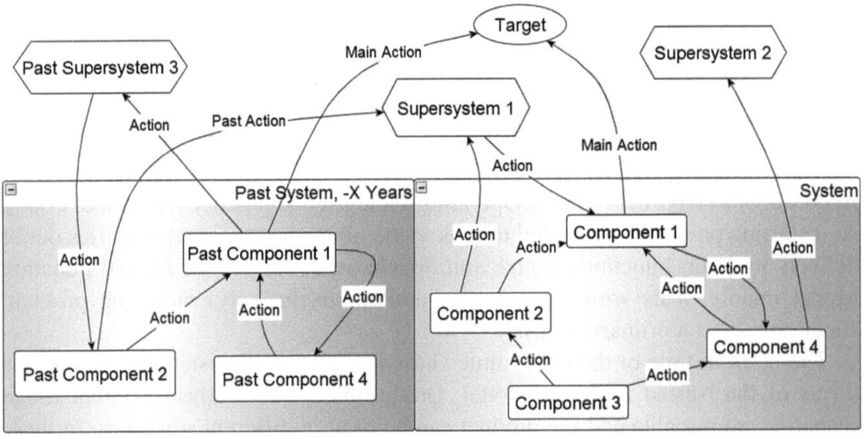

Fig. 5 Function model of past and present system. Changes: Past supersystem 3 is not existent today, component 3 has been added, interactions between component 4 and supersystem 2 and between supersystem 1 and component 1 have been added

For the horizontal aspect of each "box" of the Nested Function Model, we can define several key points in time representing significant development stages of the component evaluated. If modeled on all hierarchical levels, these points in time automatically synchronize throughout the hierarchical structure of the Nested Function Model, as past components on one system level consisted of past subcomponents on a lower system and at the same time supersystems for the component have been past versions of the supersystems today (Souchkov 2014).

For each component, future considerations and scenarios can be evaluated. Future scenarios of subsystems automatically give us clues for the system looked at, so hierarchical levels cross-feed each other with information about the development history.

Trimming expensive components or components connected to disadvantages can explicitly be used to increase the ideality of the system or each of the subsystems. With a Nested Function Model, we automatically have a vast amount of new possible function carriers and resources from the system and supersystem levels at hand.

5 What About the Main Parameters of Value?

For each present and past system, component, subcomponent (. . .etc.), we can also identify Main Parameters of Value (MPV) that have driven the development. A list of important MPVs can be attached to each "box" of a Nested Function Model. Furthermore, the main functions throughout the Nested Function Model give indications of what actually are the MPVs for each component and which subcomponents are influencing the MPVs on a higher system level. If the MPVs are quantified over the past versions of the systems or components, it might even resemble an S-Curve development that can be assessed for each component.

6 Connecting the Trends of Engineering Systems Evolution

The Trends of Engineering System Evolution (TESE) are a powerful knowledge base giving an invaluable amount of known and recurring patterns of the evolution of system characteristics. The thesis is that these patterns are objective and true for every engineering system, so they apply to a car as well as to a bolt inside a car (Training Material MATRIZ 2006, 2010).

If this thesis is true, we can apply the TESE and their mechanisms on each hierarchical level of the Nested Function Model. Having modeled past versions of the system, its components, and subcomponents (as described in 4), we can easily compare the systems or components' characteristics with the characteristics given by the TESE. This automatically gives us possible concept ideas for the future, as we can apply characteristics logically following past developments (e.g., Mono–

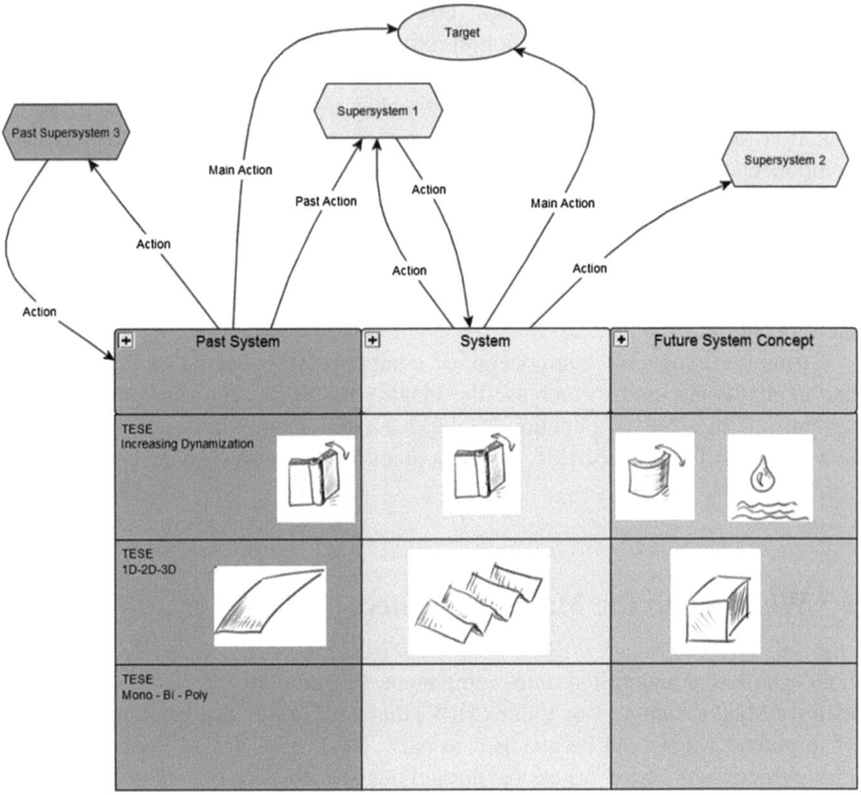

Fig. 6 Identification of TESE mechanisms. This process can be repeated throughout the nested function model for each subsystem

Bi–Poly, 1D–2D–3D...) or characteristics that have not shown up in the past. Additionally it can be evaluated, in which way the change in characteristic (or "step" of the Trend) increased the MPV. This change will most likely be quantifiable. For example, the step from a multi-hinged wiper blade of a windscreen wiper to the current flexible flat-wiper-blades leads to a more even pressure distribution between blade and windscreen, increasing the MPV "Amount of water left on windscreen after one wipe." So going from a system with multiple flexible links to a completely flexible system and thus following the Trend of Increased Dynamization leads to a quantifiable change in a system MPV (Fig. 6).

7 Case Study Paper Punch

When building Nested Function Models, it is not necessary to formulate new guidelines as all basic rules for TRIZ Function Modeling apply unchanged. Nevertheless, there are some aspects that are worth to be discussed.

7.1 System Structure Modeled as Nested Function Model

For a simple case study, a paper punch is modeled exemplarily. The focus shall be on incremental development, so we conduct a more detailed analysis of the subsystems and limit supersystem assessment to the parts, which directly interact with the punch. The starting point is the whole system "paper punch" (Fig. 7) with its interactions with the supersystems paper (target), paper offcut, table, and user (Fig. 8).

On the next hierarchical level, several subassemblies can be identified, such as base assembly, base cover assembly, punching rod assembly, and handle assembly. If the "paper punch" is considered as system level for modeling, those would be the appropriate components for the Function Model (see Figs. 8 and 9).

However, the subassemblies themselves consist of several single parts. If we set system level to be "base cover assembly," the components of the system would be guide, guide rail, and base cover sleeve. Supersystems would then not only be paper, table, and user, but also the other assemblies like punching rod assembly, base assembly, and handle assembly (see Fig. 10).

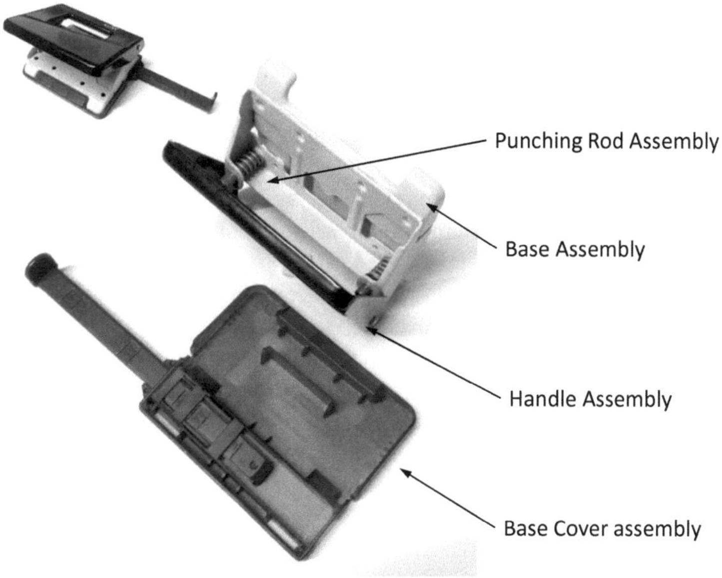

Fig. 7 Subassemblies of paper punch

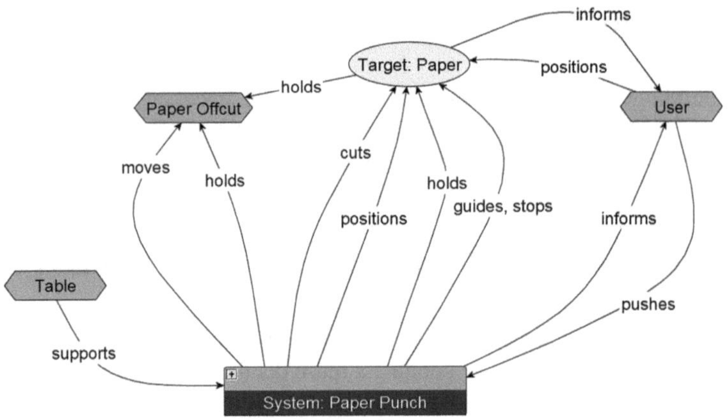

Fig. 8 Starting point on level of product/device, e.g., paper punch

The Nested Function Model can now be analyzed in different ways. One possibility is tracing the components that are involved in carrying out the main function. For this, an assessment of the model regarding all functions that are leading to the target is carried out (Fig. 11).

7.2 Enhancing the NFM with the Time Axis

As a predecessor of the paper punch, an "historic" version of the product is assessed. The time span between the products is approximately 40 years. The old system consists of the following components: Handle Assembly, Base Assembly, Mounting Block, Spring, Punching Rod (see Fig. 12).

The Nested Function Model for the old punch might look something like shown in Fig. 13.

Comparing those two function models side by side shows the changes between the old system and the new system. Exemplarily, the handle assembly is chosen for this article (Fig. 14).

It can be seen that the actuator bracket has been trimmed and its function to hold/move the punching rod has been transferred to the handle plate. Further comparison shows that the punching rod has developed into a more complex system where the spring has been combined with each punching rod. In TRIZ terms, the punching rod acquired components of its supersystem (the spring) and has developed into a more integrated subsystem of the punch.

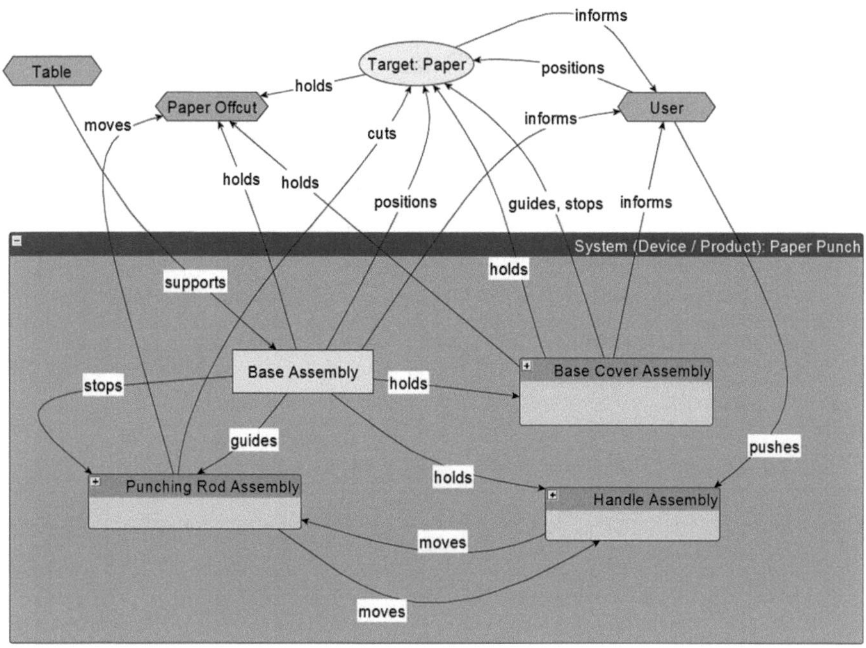

Fig. 9 Function model for paper punch

7.3 Using Trends of Engineering System Evolution to Develop Concepts for Future Products

The Nested Function Model enables us to assess changes made from one system generation to the other, simplifying the task of using TESE to develop new product concepts. When we look closer at the base plate, we realize that some components have been trimmed as a consequence of using new materials; e.g., instead of the rigid wooden base plate an elastic material is used which is able to wrap around the base plate and hold the offcuts securely in their compartment. Thus, the offcut cover, screws, and gasket could be eliminated. This resembles a clear link to the Trends of Engineering Systems Evolution, as the Trend of increased dynamization applies here (Fig. 15).

The trends then can be used for idea generation. In case of the base cover, the use of an electric field to hold/stop the Offcut comes to mind when following the Trend of Increasing Dynamization. Development stages in between might consist of material with increased elasticity like rubber or silicone that wraps around the base plate like a glove and keeps the offcut enclosed even better.

Another application of a Trend can be observed at the component that directly carries out the main function: The punching rod. Basically, the cutting end of the rod has not changed much; however, the carving on the old rod is shallow while the new rod sports a deeper carving, providing a sharper edge for cutting through the

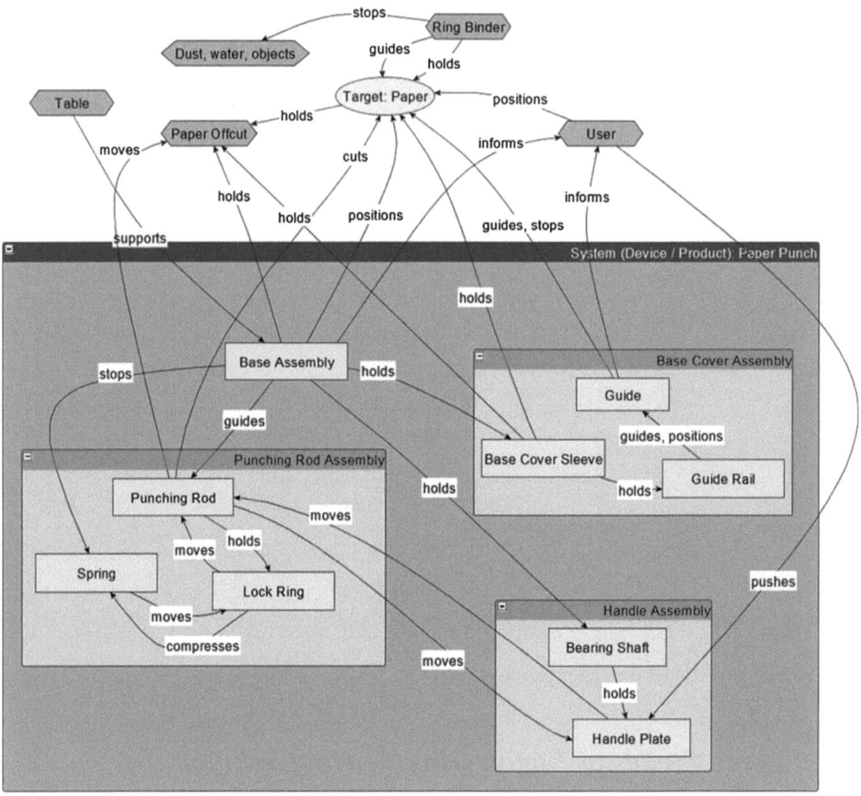

Fig. 10 Nested function model for paper punch

paper. In TRIZ terms, this represents a higher degree of dimensionality within the trend of 1D–2D–3D. A lot of possible variations to develop a more efficient shape come to mind; e.g., corrugated edges, twisted blade edges, or more asymmetric shapes could be assessed for their ability to cut through paper more effectively. Also, the movement of the rod currently happens only along a line (linear up–down movement). To follow the trend, a rotating movement combined with the up–down movement could help with the cutting process, maybe combined with a complementing shape of the cutting end of the punching rod (Fig. 16).

8 Disruptive Innovations: Influencing Factors in 9-Screen Thinking

When using the Nested Function Models for disruptive innovations, the supersystems play a key role. By taking more supersystems into account and assessing their interactions, the system "paper punch" is put into perspective and we are able to

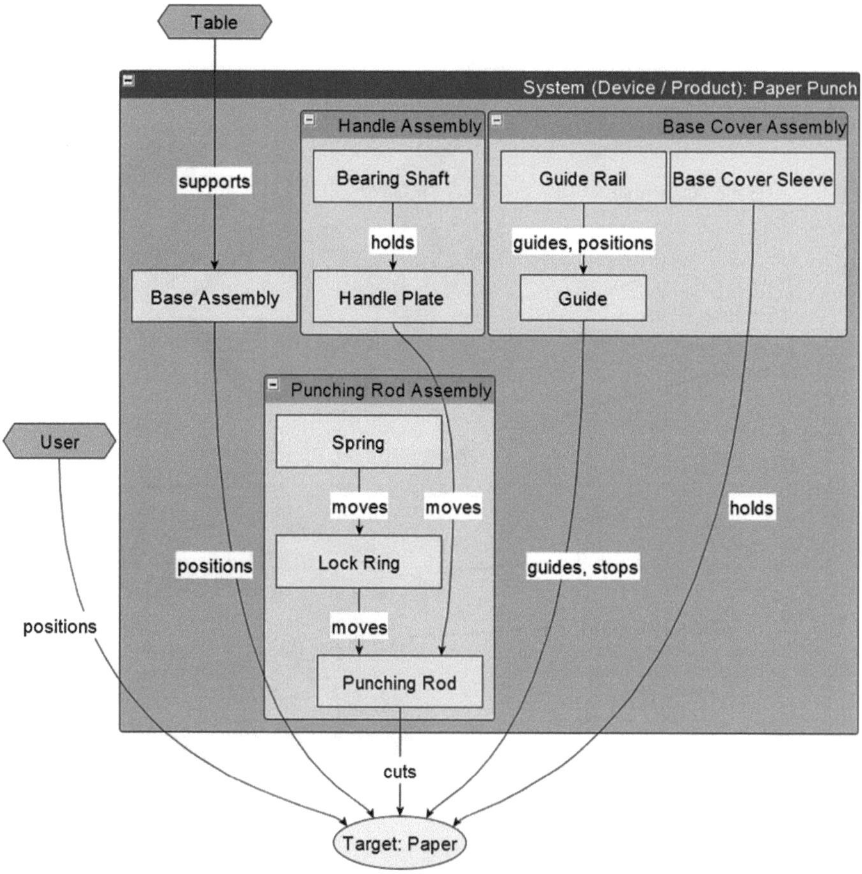

Fig. 11 Identifying the components involved in the main function

analyze the bigger picture in which the system is embedded. Globalization and the radical digitization of data already have led to several predictions, according to which the paper should have been extinct already years ago, rendering "paper punches" useless. However, this was not the case, still paper is all around and they are still being sold together with binders, folders, staplers, and so on. Nevertheless, for a strategic acting, long-term thinking company, the digitization combined with cloud services and the resulting ease of availability of data without paper should clearly be a concern. Paperless archiving and the acceptance of digital signatures are only a few supersystem changes that influence the use of the product "paper punch." According to the Trend of Increased Degree of Trimming that leads to more ideal systems paints a picture of a paperless and paper punch less world of the future. Obviously, companies that produce office supplies already have branched out into the range of digital products.

A simplified representation of extended consideration of supersystems for the paper punch is shown in Fig. 17.

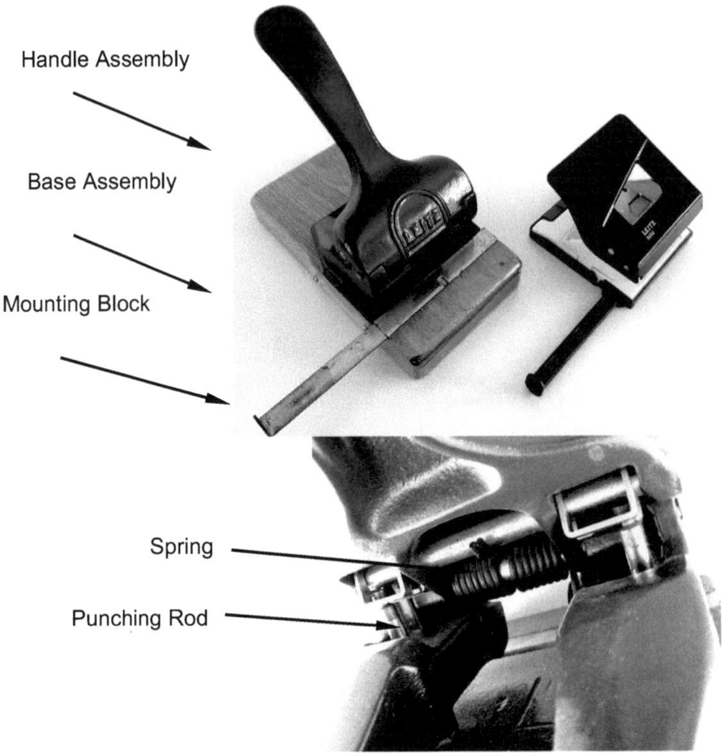

Fig. 12 Paper punch, actual system, and predecessor

With this model, several future scenarios could be developed. One of the key tools of TRIZ is trimming, which in itself is a Trend of Engineering System Evolution as well. One very important task is to assess the possibility of trimming the own product, here the paper punch. The three basic rules of trimming bring up immediate scenarios for the future while corresponding with the Trend of Increasing Ideality:

1. The punch can be trimmed if the paper is trimmed:
   The most disruptive trimming rule when applied on the product level. As described above, more and more institutions and companies try to go paperless. This results in complete new requirements and strategic reorientation of punch-producing companies.
2. The punch can be trimmed if the object performs the action on itself:
   This trimming rule sparks ideas like pre-punched paper or paper with predetermined breaking points where the pins of a binder can punch through.
3. The function can be trimmed if a different component can perform the function:
   As one option the function might be transferred to the binder, as it already interacts with the paper.

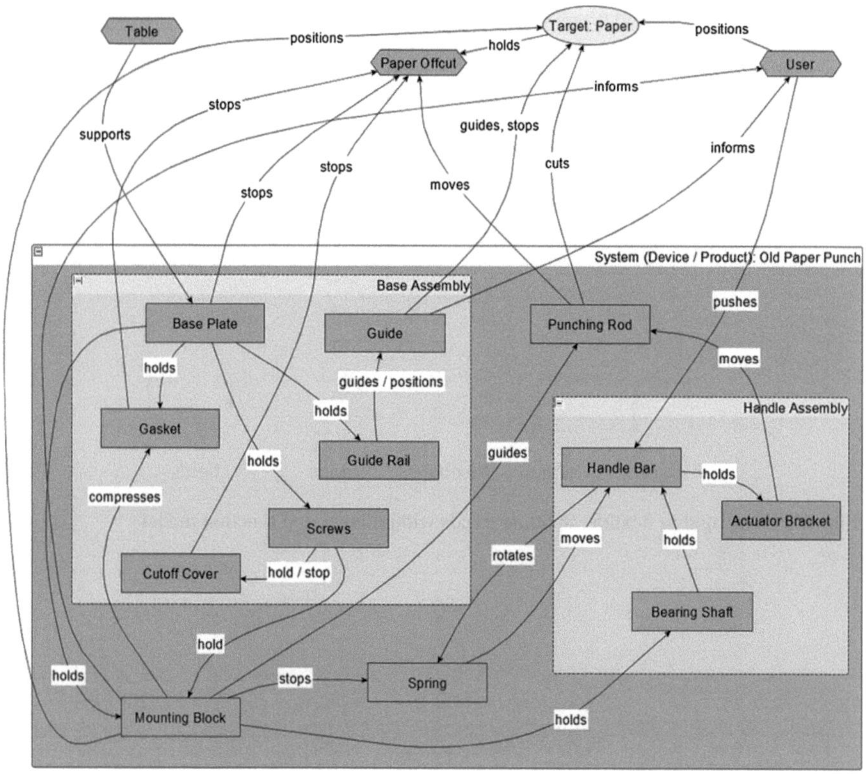

Fig. 13 Nested function model of the old punch

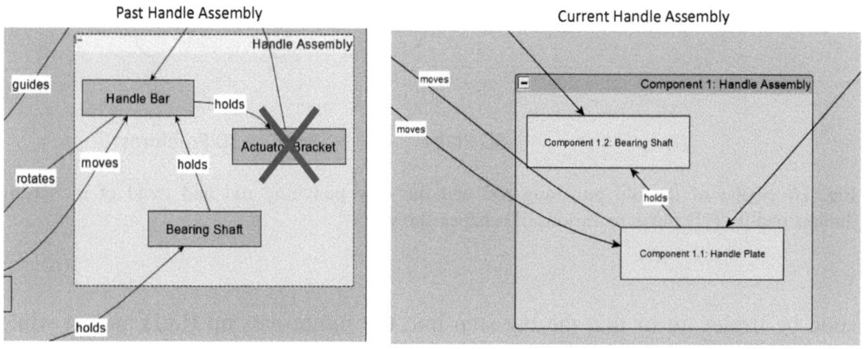

Fig. 14 Comparison of old and new handle assembly

Of course, the supersystem analysis might as well be conducted for the past system so that supersystem changes are traceable. When supersystem changes as well as subsystem developments are continuously monitored, it is less likely that developments on subsystem level are pushed despite the fact that the system might

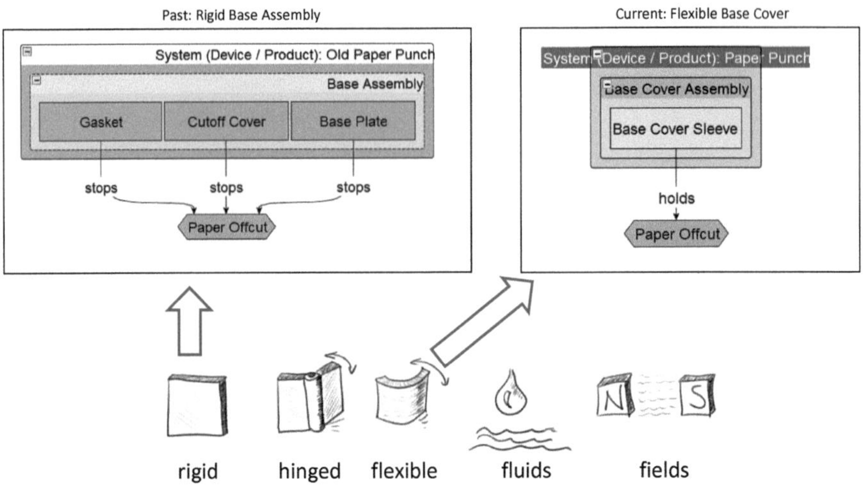

Fig. 15 From rigid to flexible. Spotting trends within the nested function model

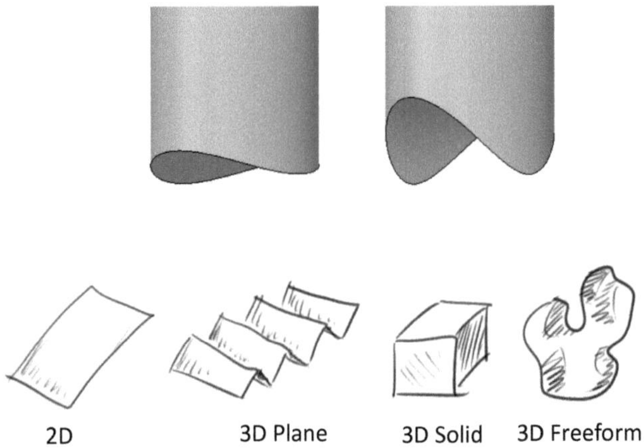

Fig. 16 Shape of the old punching rod and the new punching rod and trend of increasing dimensionality (2D plane to various 3D shapes shown)

soon be irrelevant or that the big step into the future eats up R&D budget while forgetting about incremental improvements to harvest profit on the way. Nested Function Models are a tool to support those decisions.

When all of the abovementioned aspects are brought together, the resulting TRIZ Product Map can be very complex and contains a lot of information. For the sake of the article, the authors present a simplified schematic Map for the punch example. As a frame, the typical 9-Screen Scheme is used (Fig. 18).

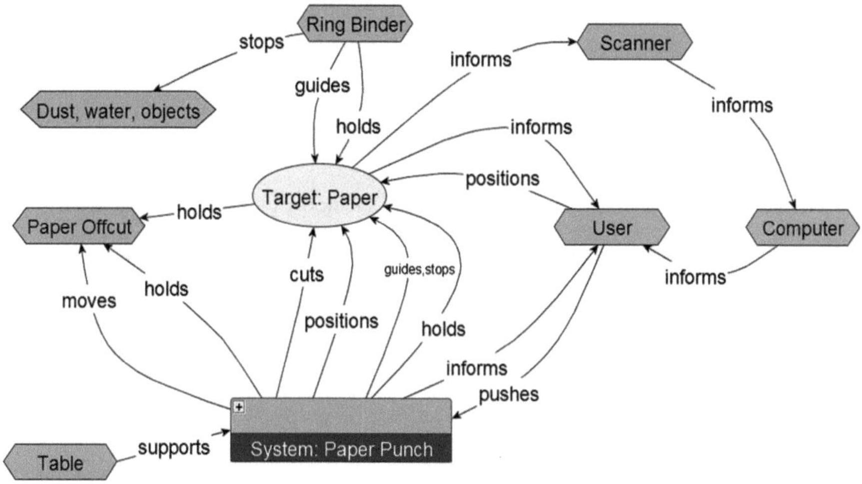

Fig. 17 Paper punch embedded into supersystems

	Past	Present	Future		
Supersystems	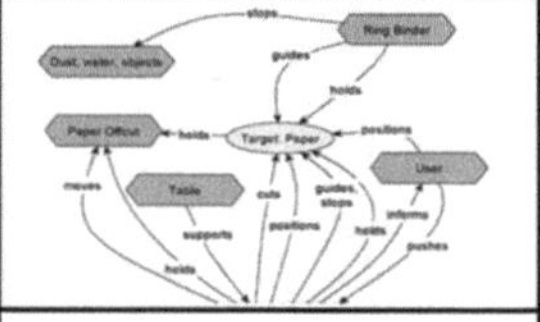		Scenario 1: Punch still needed, improved convenience for user	Scenario 2: Punch trimmed	Scenario 3: Paper trimmed
System	Paper Punch „old" 	Paper Punch „today" 	Concept 1: „Easy Punch", less manual force neede	Concept 2: Pretreated paper that is punched by binder	Concept 3: Cloud storage for important documents with digital signature
Subsystems	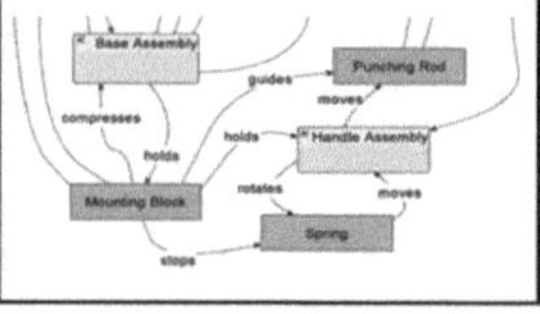	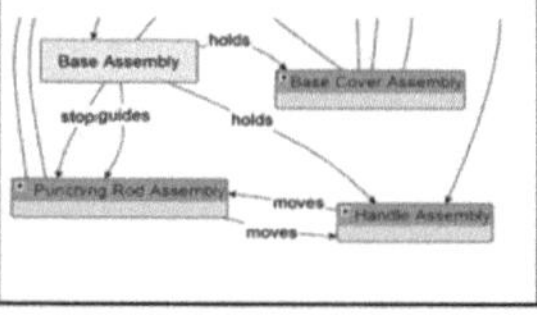	Concept 1: Improved punching rod shape	Concept 2: Multiple perforation, paper with locally adjusted thickness	Concept 3: Harddrives, digital signature and validation process, wifi connections

Fig. 18 TRIZ Product Map (simplified) for paper punch

9 Conclusion

TRIZ is often described as a "toolbox" that enhances inventive problem solving. Usually, the TRIZ tools are punctually applied when problems or limitations arise. The strategic aspects of TRIZ as the science that describes the development of engineering systems are usually separated from the operative level. This chapter connects both approaches by combining several classical TRIZ approaches and tools to support strategic product development. The proposed Nested Function Model with its expansion through the 9-Screen Model, S-Curve Analysis, and TESE is a possibility to systematically drive the development of complex engineering systems. It bridges the strategic findings of TRIZ with the operative tools for problem solving to enable a systemic development process for complex engineering systems.

The advantages of the presented Enhanced Nested Function Models are:

1. A complete functional and historical map of a product can be created.
   The product map accelerates identification of inventive and innovative potential through all system hierarchies. S-Curves and TESE Assessment provide guidelines and suggestions for future development activities and planning of new product generations.
2. TRIZ-based problem solving can be quickly initiated on each system level, because the function models are already at hand.
   The impact of changes on subsystem levels can be assessed, and development activities can aim at an even development of the system.
3. A common and transparent understanding of the whole system is created between all people involved.
   Bottlenecks resulting from functional disadvantages of subcomponents can be objectively identified and targeted. This accelerates the identification of profitable alliances, acquisitions, or potential for internal development.
   The functional oriented language of TRIZ especially simplifies the communication with suppliers concerning formulation of requirements.
4. Communication across departments (and thus "owners" of Subfunction Models) is objectified as the connection between them can be tracked down to functions connecting their respective subsystems.
   Operational activities in research and development departments can be structured and planned more systemically, as interconnections between subsystems become more transparent.
   Necessary changes, e.g., closing of production facilities due to upcoming technological changes or a shift in the company's expertise, can be planned and communicated early in advance.

All of the abovementioned aspects serve the purpose of actively shaping a companies' future by using TRIZ not only as a problem-solving toolbox, but to exploit the Body of Knowledge that TRIZ provides as an Innovation Methodology. Actively driving change by using TRIZ knowledge is therefore an important building block for a strategic corporate innovation and change management.

References

Altschuller G. S. (1998). *Erfinden Wege zur Lösung technischer Probleme*, Hrsg. Prof. Dr. Martin G. Möhrle. Cottbus: PI.

Altshuller, G. S. (2000). *The innovation algorithm* (L. Shulyak & S. Rodman, Trans.). TIC.

Cavallucci, D., Fuhlhaber, S., & Riwan, A. (2014, October 29–31). Assisting decisions in inventive design of complex engineering systems. In *Proceedings of the TRIZ future conference 2014 global innovation convention, Ecole Polytechnique Fédérale de Lausanne, Lausanne.*

Hentschel, C., Gundlach, C., & Nähler, H. (2010). *TRIZ innovation mit system.* Munich: Carl Hanser.

Koltze, K., & Souchkov, V. (2011). *Systematische innovation.* Munich: Carls Hanser.

Nähler, H., Reiffenscheid, C., & Schnittker, F. (2012). *TRIZ IT! Der TRIZ Leitfaden.* Westport, CT: Terex.

Souchkov, V. (2014, September 4–6). Multi-screen analysis for innovation roadmapping. In T. Kässi & V. Souchkov (Eds.), *Proceedings of the 10th international conference TRIZfest 2014, Czech Association of Scientific and Technical Societies (CSVTS), Prague.*

Training Material TRIZ Level 3 Certification Training MATRIZ 2006, 2010.

Taming Complex Problems by Systematic Innovation

Claudia Hentschel and Alexander Czinki

Abstract Problems are as old as mankind but a rather young field of science when it comes to categorizing, solving, or even dissolving them. The selection of a problem-solving strategy and the selection of problem-solving tools should match the problem type. Taking these aspects into consideration will not only allow to identify suitable solutions—it will also ensure a high efficiency during the process of problem-solving in general. The authors suggest a suitable model of problem types and reveal a pragmatic process of selecting adequate TRIZ tools. This structured approach is suggested, even though solving complex problems in innovation requires the amplification of creativity and inventiveness and therefore freedom. Some major TRIZ tools are assigned to problem types, not only to satisfy the often cited scientific desire to tackle problems at their correct level but with the major focus on generating a practical guideline for real-world problem-solving.

Keywords Complexity • Innovation • Problem types • Problem-solving process • TRIZ

1 Introduction

Have you ever offered a solution to an assumed problem and were not at all satisfied with the later result? For example, you have spent a lot of time and energy on designing and producing a product that the customer later completely neglected? Maybe you wanted to provide a desired product or process feature and ended up having a nonviable solution? Or were you asked to develop a better information system for your coworkers, well-knowing that more data is not automatically leading to the better? If you know such situations, you were probably facing complex problems. Perhaps you may not name them so, but you have met them

C. Hentschel (✉)
University of Applied Sciences HTW Berlin, Treskowallee 8, 10318 Berlin, Germany
e-mail: claudia.hentschel@htw-berlin.de

A. Czinki
University of Applied Sciences Aschaffenburg, Würzburgerstr. 45, 63743 Aschaffenburg, Germany

L. Chechurin (ed.), *Research and Practice on the Theory of Inventive Problem Solving (TRIZ)*, DOI 10.1007/978-3-319-31782-3_5

before. When problems are complex, they require more time, effort, and knowledge in order to be solved. This is especially true if they are not detected and treated correctly. Is there a way to get along with complex problems when sufficient knowledge, data, and time are not available? Can we solve such problems in an acceptable time span? Are we aware of the fact that complex problems are better to be dissolved rather than to be solved? Are we keeping complexity at a "healthy" level? Complexity is like a good meal: too much of it destroys all the pleasure.

Complexity can be enjoyable, as long as it is neither too confusing nor too dumb simple. This contribution tries to address what complex problems are and thins out the various taxonomies that can be found in literature with the goal to define some pragmatic baseline with a special focus on complexity in the context of innovation management.

2 What's the Problem with Problems?

2.1 *Problems in General*

Innovation is about identifying relevant problems, generating ideas on how to solve the problems, realizing a solution, and communicating its attractiveness to customers. But very often, even the first step in this process (identifying relevant problems) already causes confusion. Often it seems to be undoubtedly clear, what the problem to be solved really is. And this is where the problem with problem-solving starts.

Two things crossed the authors' minds, before they even tried to get on with problem-solving. First of all, the question "Where do problems come from?" is rarely asked. We recognize problems and implicitly know that they need to be solved, but we are often not aware of where they come from. What generates problems? Mainly they come from us being goal-oriented: when we wish to achieve a certain situation which differs from the initial situation—we call it a problem (Funke 2003). If resources are scarce and our problem-solving capabilities are low, problems easily can become severe.

Secondly, the term "problem" is obviously perceived very differently in different societies and cultures. In many cultural contexts, the word "problem" might also refer to a mainly logical or mathematical context, e.g., $1+2=?$. As a result, the term "problem" has an (emotional-wise) neutral status.

In some countries/cultures, problems are associated with a solely negative connotation. Many people even frown whenever they hear the word "problem" as such. They perceive it as a bad thing—something to be removed. However, in other countries/cultures, the term "solving a problem" is associated with terms such as "closing a gap," "meeting a need," "overcoming difficulties," and "making something work better" (Isaksen et al. 2011).

Some cultures even assign neither a positive nor a negative quality to the word "problem" but stress the openness of the outcome. In Chinese, for example, the word for problem, e.g., *wenti*, rather addresses "the question that is risen" and does not clearly distinguish between positive "chance" or negative "problem." One might also speak of an opportunity, which highlights a more positive and optimistic attitude.

Chance, opportunity, or no matter what it is called, the term problem mostly refers to a situation that is difficult to deal with. For this paper, the authors speak of a problem, whenever there is a gap between an initial and a desired situation (Sell and Schimweg 2002). Problem-solving is regarded as the process of closing the gap between "what is" and "what should be." Thus problem-solving includes all jobs and tasks to be done in order to transform an initial into a desired situation.

Interesting enough, a problem can be solved—according to the definition—not only by transforming the initial into the desired situation but also by adjusting the desired situation to the existing situation. This should be kept in mind, when we have to deal with problems that seem unsolvable to us: perhaps our target is not realistic? Or our claims are too exaggerated? We should keep in mind that problems are man-made and that we should not fall instantly into depression when we face "big" problems. Rather, we should ask ourselves how they arise and classify them, if we want them to disappear or at least want them to become more manageable. And in some situations, we should rather talk about a convergence toward a desired situation, as not only initial situations but also targets are usually highly fuzzy—especially in a world of rising complexity.

For the authors—both frequently dealing with innovation management—a structured approach to problem-solving is particularly beneficial in cases where the initial situation is hard to understand or when solutions are unclear, not available, or did not work out at first hand. Understanding structured problem-solving requires the understanding of some major fundamentals of problem theory. Therefore, the following paragraphs will deal with both, some classical and also more recent sources on complexity and complex problems, alike.

2.2 *Complex Does Not Equal Complicated*

Literature mainly distinguishes between simple and complex problems. In most publications, simple problems refer to a clear initial situation and a known way to get to a solution. An example for a simple problem is: What is $1+2=?$, which for most people would constitute rather a mathematical problem or task, but not a problem in the sense that it would be hard to solve. But with the eyes of a toddler or child before going to school, you might see it differently: unless the reading of signs and numbers, the meaning of "+," and the logic behind addition is understood, the problem remains unsolved. So, obviously, simple problems can also be difficult, depending on the knowledge of the person. Also things, looking simple at first glance, can be difficult at the second (Norman 2011): take chop sticks, the proper use of which requires—among other things—even the knowledge of culture and customs. Is driving simple? It depends—on the driver, on the situation, and on the driving education.

So it seems that problems difficult to solve can more easily be transformed to some sort of a solution by a previous learning process (Sell and Schimweg 2002). The repetition of the underlying algorithm for solving similar problems creates practice and leads to a problem-solving capability for problems of the same and similar kind.

That is good news, and especially true for well-structured problems. However, the focus of this contribution is more on the type of problems, where knowledge alone does not directly deliver sufficient results or even provides no solutions at all. For the authors, the starting question was how to call such problems and then ask further how to allow or even facilitate the selection of adequate problem-solving tools later on. Our first guess was that the term complex problem would lead to deeper insights.

In an attempt to answer the starting question, dictionaries were not helpful, as they often do not even distinguish between complex and complicated problems. Both terms are emphasizing things with many intricate and interrelated parts and aspects. If one does like to distinguish between them, one can only do so in the extremes: take a recognizable pile of sand. One would presumably tend to call the arrangement complex. Taking away one grain after another will obviously reduce the complexity. But, when would you stop calling it to be complex?

Or take the design of a machine with many mechanical and electrical components. Even if you know all the parts and their relations, a faulty switching circuit can lead to serious damages, and maybe you do not know right away what the proper remedy would be (Zeichen 2014). Despite the fact that the elements and their connectedness can fully be enumerated, the system might behave in a weird and unexpected way. While this is part of the engineers' world, still some authors claim that once all parts and relations are known, one should "only" speak of a complicated system. But when would you call an airplane or a space shuttle complicated rather than complex? Probably, it is the randomness of the system's behavior that could quickly make one perceive a system as a complex one. So while complication seems to be more in the eyes of the beholder, complexity is part of the world—and has always been; there is no sharp distinguishing demarcation line between the two terms (Holland 2014), and thus there is no sharp and generally accepted dissociation between the two terms in literature either. In the middle-ground, the borders stay blurry.

3 Complex Problems and Innovation

3.1 Complex Problems in General

Complexity and complex problems have been addressed by a number of research fields and by many well-known authors, beginning in the 1930s. Such authors are, e.g., the Germans Zwicky (1966), Rittel (2013), and Dörner (1989) or the Americans Forrester (2013), Ackoff (1974), or Ritchey (2011a, b), all of them seeing problems occurring in different levels with the extremes stretching from "puzzle" to "mess."

The above mentioned books call a situation where a well-defined issue is to be tackled and a well-defined solution is available (which "merely" has to be worked out), a "puzzle": a given number of defined parts with a clearly defined relation to each other need to be put together to a specific solution. If there are only few parts, this may be covered by routine, which basically is a question of time and patience.

Besides the terms already being introduced, classical authors also use further categories while classifying problems. One of these classifications is called a "messy problem," a culmination of thinking approaches to a set of problems that won't be solved by a simple, narrow focus.

So, a "mess" climbs the highest level within the range of complex problem modeling and solving literature. A synonymous and recently more frequently used word for it is "wicked problem" (Ritchey 2013; Buchanan 1992; Lindberg et al. 2012). The term "wicked" was originally coined in 1972 by Rittel (2013), who used the term "evil," if one translates the German word "*bösartig*" correctly. More recent literature uses "wicked" not so much in the sense of evilness, but it could lead to unintended consequences, reaction, and even back-fighting, once one tries to overcome it (again Ritchey 2011b). Wicked problems or messes address highly complex, unclear issues that are not at all structured or formed. When you have a mess, it is even unclear or you do not even know what the problem *really* is. This is the case with, e.g., administrative or political situations, when you are trying to defeat poverty or drug addiction or boost health and education.

Again, the entire space is not well defined, but sources agree upon that in general complex problems occur, when components, their behavior, and relationships are changeable and are not clear enough to be named by the problem solver. Some characteristics of complex problems are enumerated (Funke 2003, p. 126 ff.) as follows:

- Indeterminacy of problem situation
- Connectivity of variables
- Dynamics of problem situation
- Intransparency of involved parameters and objectives
- Polytely (i.e., many or even unknown number of objectives to be observed or met simultaneously)

Other publications enumerate even more characteristics, e.g., Rittel, Buchanan, and later Ritchey name 10, with missing stopping rules for problem solvers and uniqueness, to name only two of them (again Ritchey 2013). As a consequence the behavior of such systems varies, is manifold, and is even fraught with ambiguity. Therefore, no objectivity exists (Schönwandt et al. 2013). Any approach to name or even define the problem or solution is a subjective one. Such systems seem unmanageable, even with lots of effort and lots of knowledge. For such problems, there is no clear "right" or "wrong" either, no "correct" solution to be focused on. The longer we look at the situation, the more solutions even seem possible. Any place or situation, where people and their human behavior play a major but unforeseeable role, all social, commercial, and/or organizational systems, briefly: all systems that involve people—the most complex and adaptive systems known—more or less imply complex problems.

Anyhow, we have to deliver a solution, often under time and many other—ever rising—constraints. As the authors assume that more and more such complex problems call for solutions, this is the reason why they try to address it herewith a little deeper.

3.2 Complex Problems: A Practicable Approach

When it comes to innovation, managers are probably less in control of things than they imagine. Business situations, in which new products, processes, or services need to be defined such that the user or customer really wants them and thus is also willing to pay for, give more than enough room for complexity. Also finding new application fields for an invention, e.g., when a famous glass producer has found a way to produce glass tubes with alveolar cross sections but has no idea who would need and apply such a competence, he automatically faces a complex situation.

A similar innovation-related problem occurs, when companies try to extend the fields of application for their products and technologies into new market fields. The same is true for a situation in which a new innovation management process is established, and its results do not match with the stakeholders' needs or that the stakeholders' needs have evolved into another previously not "foreseen" direction.

Problems in innovation management usually address more than just a single issue. Usually it is rather a set of interrelated elements, a number of parallel existing concerns that interact in a way that is hard to predict. The authors have carried out an in-depth literature research to get a grasp on the variety of problem distinctions in terms of innovation issues, hoping to come up with definitions that do not contradict other fields of problem-solving literature. Figure 1 represents, what the authors suggest as a common ground, linking the different interpretations of problem types and their main characteristics.

The chart spans a three-dimensional space which allows assigning problems into the following types: chaotic, complex, complicated, and simple problems. Applying this frequently used classification allows the authors to later implement

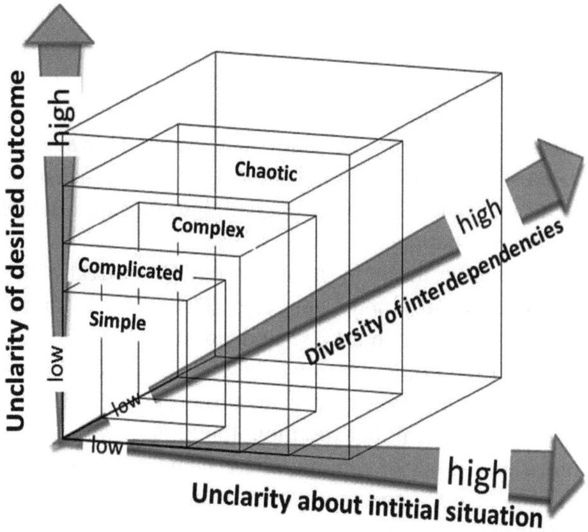

Fig. 1 Problem types and their main indicators

elements of problem type definitions and solving strategies of other sources such as the Cynefin Framework (Snowden 2000).

The authors suggest that a given problem has to be assigned into the next higher class whenever one of the three aspects—represented by the chart's dimensions—exceeds the characteristics of the current class. One dimension tending to a "higher" class is—according to the authors' perspective—perfectly sufficient to catapult a problem to a higher problem level than classical literature would suggest.

The dimensions that allow our taxonomy are threefold: *the first dimension* addresses the initial situation, where the ability of naming and understanding all elements of the issue to be tackled rather suggests a simple problem at hand. The more the initial situation is unclear, the more the problem leaves the level of simplicity, even if there seems a solution to be at hand. The *second dimension* refers to the knowledge about a solution. Most sources suggest that there are problems where a solution is known and the other where no solution comes to mind. Once we have to tackle the latter type of problem, it has for the authors lost its "simple" status. The *third dimension* suggested is the diversity of the behavior of a system, producing unconsidered side effects and surprise when deciding for a solution. If interdependencies are severe and manifold and their impact varies, we would suggest ranking any problem—even an extremely well-defined one—into a higher class than the initial situation may suggest alone. With this taxonomy in mind, four different problem types can be distinguished.

Chaotic problems are those where uncertainty, disorder, the unknown, and randomness dominate the system behavior. A chaotic system seems to produce noise without sense and displays utmost dynamics where no signal or pattern can be derived. Taleb (2012, p. 13) enumerates 16 members of the chaos family ranging from uncertainty to disorder up to the unknown. Chaotic problems are random and unpredictable, and even if the nature of a system is deterministic, this does not necessarily make it predictable for us humans. It is important to notice: being chaotic is not necessarily a negative quality as systems can either harm or gain from being exposed to this volatility.

This problem type is rather treated for the sake of completeness in this paper; it does not constitute the main focus here. However, it is considered as the most interesting and challenging subject for future investigations. Herewith, the focus is laid on another, also very challenging, question, which is: what to do in complex problem situations?

In *complex* problem situations, one has to deal with phenomenological situations that are not fully understandable. They refer to things such as society (Luhmann 1987), economy, markets, and cultural behavior as well as leadership (again Snowden 2000), to name a few. They are man-made, but develop on their own to reach some kind of self-organization (again Taleb 2012, p. 56). “They may not be strictly biological, but resemble the biological in that, in a way, they multiply and replicate—think of rumors, ideas, technology and businesses” (again Taleb 2012, p. 56). Taleb suggests a cat to depict such systems, and others (again Snowden 2000) depict a frog for visualizing complex systems. Anyhow, such systems and problems rather do not resemble mechanical ones.

Complicated problems are not complex problems: complicated problems are discoverable, more predictable, and less dynamic in their behavior. If a person has enough knowledge, sufficient technical background, and can apply enough effort or other resources, e.g., can call in an expert, this type of problem can be solved by analysis and expertise. The question is not whether it is possible to analyze and understand the problem. The question is rather how and where to get the required expertise and resources from and whether the possible benefits justify the required efforts to do so.

Simple problems are well-defined issues with well-defined and notable initial parameters. The desired outcome is clearly defined. Simple problems follow a strict logic—they consist of clearly recognizable and constant components, whose interdependencies and behavior can be defined and predicted.

Weibel enumerates the following examples for each category: baking a cake is a simple problem. Sending a rocket to the moon is a complicated problem. Raising a child is a complex problem, and climate and weather change is on the edge of chaos (Weibel 2014). Many scientists already suggested that a number of complicated problems can be reduced to a number of simple problems.

Astonishing enough is that the other way around is equally true: a seemingly simple problem in terms of its initial situation and desired outcome may quickly turn into a complicated, complex, or even chaotic problem. Take a simple problem: the higher, e.g., the number of elements and their interconnectedness become, the more it becomes complicated. If the roof of your house is leaking and you do not manage to model the interconnected parameters that obviously interact, such as weather, rain, thermodynamic behavior of air, humidity, material pairing, the influence of temperature, and so on, you may understand what technicians are dealing with when they do not find the “right” solution, even if the desired outcome of a watertight roof is clear enough.

Also situations, where the initial situation is clearly defined, and there is some knowledge about how the variables interact, the lacking solution or unclear solution space may suggest a complex problem. Take a typical innovation problem, e.g., to design a new service: such situations do not have a single, clear-cut solution. The solutions are manifold, involving “depending on” relations, such as depending on money, on technology available, on the power of a person, and on the global economy, to name only a few possible “dependencies.”

Problems, where the desired outcome is unclear, are difficult or even seem insolvable: how to search for a desired outcome in case it is unclear and the knowledge about it isn’t there? Uncertainty concerning the desired outcome requires more than just search and/or application of knowledge. It needs a deeper dive-in and calls for a reconciliation with superordinate goals.

The aim of the authors is not to eliminate complexity, but to avoid messy or inadequate processes for solving such problems, much in the sense that it is much better to dissolve a problem than to solve it (Ackoff et al. 2010). The authors rather try to tame complex problems; otherwise, the problem (dis)solver would spend all his/her energy on endlessly analyzing the system and would—since the system cannot be fully described—never get anything done.

4 Taming Complexity: The “Golden Way” of Problem-Solving

We have to look at the nature and type of a problem before trying to solve it. And this should be done in the very early stages of design and development, well ahead of the conception phase. In many cases the analysis of problem types will lead to the conclusion that one has to deal with a problem that has complex characteristics. Complexity is neither good nor bad: it is the confusion about it that is bad. If we treat complex problems with tools that belong to other problem types (e.g., tools for complicated or simple problems), complexity will fight back.

Only when complexity is random and arbitrary, i.e., on the edge of chaos, we have reason to feel annoyed, but again it is our confusion about it that we should complain about. In that latter case, we probably cannot ensure successful problem-solving, but should reduce risk to a minimum.

It is often cited that it is a mistake to manage a mess by simply ignoring the messy part, treating the rest as a complicated problem and solve it as a puzzle, as Pidd (2009, p. 43 ff.) puts it. Nevertheless we have no other choice than trying to tame complexity by deriving underlying principles, structures, and patterns. A selection and thus reduction of the system to the “most important” elements (Dittes 2012) make complex problems easier to handle. Planning tests and learning from them in an iterative manner and even solve some parts of a complex problem as puzzles can help to manage highly complex tasks. This reduction into a few but decisive parts allows only viable solutions to pass (again, see early authors, e.g.,

Luhmann 1987, to recent authors, e.g., Taleb 2012; with Dittes 2012; and Weibel 2014, confirming this idea). This strategy does not make the problem itself less complex, but we feel less confused and get at least some chance to manage and master it. Thus the question is not whether we want to reduce/simplify reality, but rather whether we are aware of it and whether we address and manage a possible simplification successfully.

Focusing on the conception phase of a solution for a complex problem or system, a systemic approach of thinking is required. Systemic thinking—in short—means that complex challenges require an understanding of what the problem really is, a basic idea of what the main elements creating influence are, how the individual components interact and regulate themselves before even thinking about how to create adequate ideas for solutions. After all, the solution should be feasible and viable. The main idea is that each system is part of a whole and consists of subsystems (Ninck et al. 2014). So each problem should be looked at from an external perspective with a view to the problem environment, as the solution has to integrate itself into its surroundings, and—closely together—it should be looked at the system itself, as it has to be realizable and feasible. So the more you are able to zoom out, the lesser details will be visible. However, chances are that the most influencing elements and their interaction patterns that create a complex problem will become visible more easily. And the more you are able to zoom in, the more you are likely to be able to design a system that fulfills all the desired functions and sub-functions.

So the authors express the need for selecting problem-solving techniques according to the problem type at hand. Even more, it is important to provide both techniques *and* processes to efficiently solve the ever increasing complexity in modern social and technological environments.

5 The Right Problem-Solving Strategy to a Given Problem

5.1 A General Problem-Solving Model

Problem-solving is far too important in professional life to be dealt by chance only. By using a standardized problem model in combination with adequate problem-solving tools, problem-solving can be become a stable and robust process even if complex problems are involved. Subsequently, a general problem-solving model will be described and a linkage/combination between the different phases of problem-solving and typical TRIZ tools will be introduced.

5.2 *Problem Awareness*

Problem awareness is a fundamental prerequisite for problem-solving. In this context, problem awareness is regarded not only as a rough status of "...being aware that there is some kind of problem...." Problem awareness should focus on identifying the major sources causing the problems—often called "root causes." In complex situations, the identification of the root cause (the root problem) can be very challenging. Since a problem can be defined as a deviation between an initial and a desired situation, problem awareness also needs to address both aspects of a problem: it requires a clear picture of the desired outcome and also a proper identification of the initial situation.

5.3 *Problem Type Analysis and Identification*

As a next step of the overall problem-solving process, the problem type (chaotic, complex, complicated, simple) should be identified. This is a very important step, since different problem types require different strategies and tools for problem-solving. As soon as the problem type is identified, different strategies for solving the problem can be applied (Table 1). According to the preferred strategy, adequate TRIZ tools can be suggested in advance.

Chaotic Problems

In case of chaotic problems, a strategy of fast adaption is suggested. As a clear input-to-output relation is not available in chaotic systems, it is neither promising to spend a lot of energy in analyzing the system nor it is reasonable to invest time and energy into determining an optimal input to a system. In this case it is far more important to clearly sense the systems output and to adapt as fast as possible to it. Within a chaotic space, the opportunities and effectiveness for classical problem-solving are limited.

From the authors' point of view, there are no specific TRIZ tools that can be assigned to solving chaotic problems. TRIZ itself is a very logical problem-solving methodology and therefore its effectiveness automatically suffers whenever input and/or system information is incomplete. Also TRIZ is not able to change the nature of chaotic problems by actually transforming them into complex, complicated, or simple problems. Nevertheless, TRIZ can be helpful in context with chaotic problems: in a truly chaotic world, fast sensing in combination with agility and a high flexibility to quickly adapt to changes are the key criteria for success. TRIZ can be used to derive adaption strategies (which itself will usually be a complex problem). It also can enhance creativity in general and therefore give support in situations which are characterized by high uncertainty. And again, fast adapters are the winners in a truly chaotic world. TRIZ can be used to increase the adaptability of current systems toward chaotic changes.

Table 1 Problem types and their main characteristics with preferable TRIZ tool

Problem type	Cause–effect relation	Main challenges	Strategy	Crucial capabilities	Preferable selected TRIZ tools
Chaotic	Noise dominates signal -> no useful relation	High dynamics, turbulence, noise	Fast adaption	Fast sensing, agility, flexibility, creativity, etc.	Apart from generally enhanced creative thinking capabilities: none[a]
Complex	Phenomenological, not fully understandable	High number of unknown interdependencies within the systems and between the system and its environment	Work based rather on hypotheses than facts, deriving and using patterns, testing, learning	Planning and designing tests, knowledge of patterns, ability to transfer patterns to new applications	Engineering system evolution (ESE), system operator ("9 –Windows"), ideality, smart little people
Complicated	Discoverable but not immediately apparent	Identification and modeling of the most crucial interactions, distinction of systems	Focus on crucial (sub)systems, modeling, analysis	Analytical minds, expert knowledge	Flow analysis, technical contradictions, physical contradictions, substance-field analysis (SuF), trimming, feature transfer
Simple	Logical, predictable, consistent	Precise and efficient transfer of principles and solutions to the individual problem	Adapt and apply	Knowledge, accuracy, efficiency	Catalogue of effects

[a]Tools for solving chaotic problems are beyond the scope of this investigation

Complex Problems

Complex systems are—in comparison with chaotic systems—still very "fuzzy": the initial situation is difficult to sense, there is no clearly defined target situation, and a high number of interdependencies within the system itself and between the system and its environment exist. A manageable, complete, and reliable system model is not available. The system's output on a given input is difficult to forecast, and the system behavior can only be predicted on the basis of input–output patterns that have been observed in the past. Information that is not available in the form of such patterns requires exploring the system's behavior in order to gain new insights and by this means: gaining new patterns.

A typical, highly complex problem engineers and businessmen have to deal with is the question "How will a given system develop in the future?" Forecasting the development of a product is highly complex. This is true, since neither all influences are known nor the strength of the different influences on the outcome is properly understood. Questions illustrating the complex character of predicting a system's future are: What new technologies will be available in the future? How strong of an impact will they have on a certain product? How will the global economy develop and what impact will this development have on the customers' needs?

TRIZ provides several tools that address complex problems. A typical tool would be engineering system evolution (ESE). ESE makes use of development patterns of engineering systems that have been observed to be true, beyond the borders of specific fields of industry and also beyond a specific time range. These patterns help to develop products further. Interesting enough, this tool works, although the mechanisms behind the patterns used are too complex to be completely understood (Altshuller 1998; Petrov 2001).

While the ESE tool only allows applying given patterns to a problem, the multiscreen approach (MSA) or system operator (better known under the term "9 windows") even goes one step further: MSA allows generating new patterns. By extrapolating trends within the product's super- and subsystems, this tool enables the user to create a hypothesis of the product's future development. By its nature, the MSA tool ensures that the predictions made are consistent with the development trends being formulated for the corresponding super- and subsystems.

Both tools mentioned can very well be applied to reduce and handle complex technological problems and can be adapted to nontechnical complex systems as well, e.g., for finding out what the customer really wants (Ulwick 2005). However, the major strengths of the tools are clearly located in the very same domain that they have been invented for: deriving the "voice of the product" of complex technical systems.

Additionally, the tool "Smart Little People" is highly applicable in complex situations: while "playing" with imagination by making small, all-round characters taking over the role of an undefined resource, complex situations can be theoretically simulated according to their possible reaction. This helps probing hypothesis and learning from possible outcomes. In consequence, these tools do not suggest a solution, but prepare a test-and-learn field.

Complicated Problems

Complicated problems are—based on the definition of the term that was given before—problems that can be described. Thus there is the option to fully understand the system's structure based on clearly notable cause–effect relationships. What usually makes these systems (and therefore also problems related to these systems) difficult to handle is that their structure and their inherent cause–effect relationships are not instantly apparent. Thus difficult problems are requiring a detailed in-depth analysis of the related systems. TRIZ provides tools that allow the user to carry out in-depth system analysis, providing well-defined, universal, and robust system analysis processes.

Other TRIZ tools addressing this type of problem are the classical TRIZ contradiction tools, from technical to physical contradictions. Also, substance-field analysis (SuF), trimming, and feature transfer represent tools for the complicated problem domain. These tools are only applicable if an in-depth system analysis and system understanding precede the application of that tool. Here, solutions are proposed, even though their abstraction level leaves room for realization alternatives.

"Simple" Problems

As stated before, the term "simple" should not be underestimated or misunderstood as "easy." The term "simple" is purely related to the fact that the cause–effect relationships within a given problem are well understood. As a consequence, solving "simple" problems can still be a difficult and challenging task. A typical example for a "simple" task would be finding a physical principle that is able to generate a desired function, e.g., convert a temperature into electric current. TRIZ uses databases, e.g., catalogues of physical and other effects, to find new or alternative solutions for "simple" problems, which is why we find catalogue of effects in this problem domain.

6 Problem-Solving Process with the Help of TRIZ

TRIZ can be of tremendous help during the process of solving real-world problems. It has been suggested to assign specific TRIZ tools to certain problem types (mainly to complex, complicated, and simple problems). With the above suggested repartition of TRIZ tools according to problem types, the remaining question is: what are the indicators that allow the selection of the "right" TRIZ tool if one does not have the time to apply all of them? The authors' answer—that will be elaborated further—is: if probing is the prevailing aim, then the tools for complex problem-solving should be applied. If analyzing the system is prevailing, then the tools for complicated problem-solving should be chosen. And last but not least, if categorization is prevailing, the "simple" problem-solving tools should be considered.

Although the enumerated TRIZ tools above are far from being complete, some tools come in the authors' minds that are not (yet) to be clearly assigned to a special problem type. Take the "function analysis" (FA) tool. It allows the user to "zoom into" the product. It helps identifying the system's components and the network of interactions between these components, and it also allows recognizing the *type* of interactions that exist between them. As a result, complicated systems are described by a larger number of smaller subsystems/components and their—compared to the overall system—relatively simple interactions. The tool allows the user to systematically explore cause–effect relationships within systems. As soon as the system is modeled, other TRIZ tools can be applied, which allow to strengthen desirable interactions and to weaken or even eliminate nondesirable and harmful functions. With the system's inner structure lying open, TRIZ even qualifies the user to redesign the system such that existing interactions are eliminated and new interactions are generated. Thus TRIZ enables the user to decode the inner structure of systems and therefore allows a systematic optimization.

Applying FA for a certain problem level does not completely solve the corresponding problems, but rather reduces the problem level of a given problem in order to gain an advanced understanding of the problem. Thus a complex problem is, e.g., converted into either a single or a set of complicated problems which are later transformed into simple problems, which then finally are solved. This indicates that there are some sort of transition tools that help reduce a problem level.

Other tools that are not yet clearly to be assigned to are, for example, the Innovation Situation Questionnaire (ISQ) and Resources. These tools are applicable in situations where one does not (yet) know which problem space one is in. These tools prevent from tending prematurely to one of the four problem types and first ask questions and gain information about the system or problem at hand. These tools do not deliver solutions, but rather gain more problem or system understanding. The authors will address their special function and suggest an additional problem type assignment, which will be dealt with in more details in their future publications.

It is important to mention: whenever there is a problem, the question should first be what problem type it represents. Solutions can then be generated with the adequate tools that are assigned to the problem type at hand. Whenever the processing of the problem on a certain level fails, the problem-solving process needs to step backward to one of the prior levels. Thus, failing on the complicated problem level asks for returning to the complex problem level, where new directions for system development and new directions for an increased level of ideality are aspired. Failing on the simple level requires at least returning to the complicated level where alternative problem models (e.g., an alternative problem model especially suited for a newly identified principle) are generated. The problem-solving model introduced here is of a highly iterative nature, allowing to step back whenever a result of a certain step is not satisfactory.

Problem-solving could start with applying the tools that belong to the level the problem actually belongs to. Alternatively, the problem-solving could directly start by reducing the problem level and by solving the corresponding lower level right away. At the moment, the authors do suggest a general rule on where to start with problem-solving, but stress the fact that it is extremely important—and beneficial at the same time—to always be aware on what level one is on, to what advantages and disadvantages this has, and to preferably use tools that are able to treat the problems at this level.

7 Conclusion

Complexity is omnipresent in our lives. We cannot evade it, but we have to master it. Much of the effort of this paper was spent on transferring the different opinions and statements of different scientific fields covering complexity and complex problems into a current and practicable innovation context. All in all, this was more challenging than expected: definitions found in literature are hardly universal; furthermore, different science fields describe complexity in diverse manners. The authors have mainly followed an understanding of complexity based on system theory and tried to substantiate their idea of a generalized problem-solving strategy.

Identifying that there is a problem—a deviation between an initial and a desired situation—is an essential first step in problem-solving. In addition to realizing the mere existence of a problem, it is even more important to clarify which type of problem there is, as different problem types also require different strategies for solving them. As soon as the problem type is identified, a set of possible problem-solving tools can be used. The authors suggest specific problem-solving tools for specific type of problems. The tools have been selected with a particular focus on the application of TRIZ tools. However, the model itself is not limited to TRIZ tools.

In this context, TRIZ helps to reduce complexity in a step-by-step approach: transforming a complex into a complicated and transforming the complicated into a simple problem. This is happening without neglecting the initial complexity of the situation at hand, this being the reason why the authors speak of taming complex problems. The preferred tools are furthermore not strictly limited to one indicated level of problem type, but rather floating in their applicability for the next lower level. Furthermore, the model introduced here is—just as TRIZ in general—to be considered as work in progress. After all, the success of the problem-solving process in general will strongly depend on the effectiveness of the problem-solving tools used.

The authors' objective of this contribution was to generate a link between problem models on the one hand and TRIZ—as an especially powerful problem-solving methodology—on the other. They hope to contribute to more linking the different concepts of problem-solving on one side and systematic innovation on the other. A bit to their surprise, the mere work on problem types not only strengthened the ability of the authors to analyze problems more clearly but made them see disturbances and trouble—as they happen in complex situations all the time—from a new and especially from a more relaxed perspective.

References

Ackoff, R. (1974). *Redesigning the future*. New York, NY: Wiley.

Ackoff, R. L., et al. (2010). *Systems thinking for curious managers*. UK: Triarchy Press Ltd.

Altshuller, G. S. (1998). *Creativity as an exact science—The theory of the solution of inventive problems* (4th ed.). Amsterdam: Gordon & Breach Science Publishers.

Buchanan, R. (1992, Spring). Wicked problems in design thinking. *Design Issues, VIII*(2). Chicago: School of Art and Design, pp. 14–19.

Dittes, F.-M. (2012). *Komplexität—Warum die Bahn nie pünktlich ist*. Berlin: Springer Vieweg.

Dörner, D. (1989). *Die Logik des Misslingens. Strategisches Denken in komplexen Situationen*. Hamburg: Rowohlt Verlag.

Forrester, J. W. (2013). *Industrial dynamics. Reprint of 1961 edition*. Eastford: Martino Fine Books. Original Version from 1961.

Funke, J. (2003). *Problemlösendes Denken*. Stuttgart: Kohlhammer Verlag.

Holland, J. H. (2014). *Complexity. A very short Introduction*. UK: Oxford University Press.

Isaksen, S. G., Dorval, K. B., & Treffinger, D. J. (2011). *Creative approaches to problem solving* (3rd ed.). California: SAGE Publications.

Lindberg, T., Köppen, E., Rauth, I., & Meinel, C. (2012). On the perception, adoption and implementation of design thinking in the IT industry. In H. Plattner, C. Meinel, & L. Leifer (Eds.), *Design thinking research. Studying co-creation in practice. Reihe: Understanding innovation*. Berlin: Springer.

Luhmann, N. (1987). *Soziale Systeme* (1st ed.). Frankfurt am Main: Suhrkamp Verlag.

Ninck, A., Bürki, L., Hungerbühler, R., & Mühlemann, H. (2014). *Systemics. Viable solutions for complex challenges*. Leipzig: Heuris Publishing.

Norman, D. A. (2011). *Living with complexity*. Cambridge: MIT Press.

Petrov, V. (2001). *The laws of system evolution. TRIZ futures 2001*. 1st ETRIA Conference.

Pidd, M. (2009). *Tools for thinking. Modelling in management science* (3rd ed.). UK: Wiley.

Ritchey, T. (2011a). *Modelling complex socio-technical. Systems using morphological analysis*. Stockholm: Swedish Parliamentary IT Commission.

Ritchey, T. (2011b). *Wicked problems—Social messes. Decision support modelling with morphological analysis*. Berlin: Springer.

Ritchey, T. (2013). Wicked problems—Modelling social messes with morphological analysis. *Swedish Morphological Society: Acta Morphologica Generalis AMG*, *2*(1), 8p. ISSN 2001–2241.

Rittel, H. (2013). *Thinking design. Neu herausgegeben von Reuter, W. D. und Jonas, W.; Board of International Research in Design BIRD*. Basel: Birkhäuser Verlag.

Schönwandt, W. L., Voermanek, K., Utz, J., Grunau, J., & Hemberger, C. (2013). *Komplexe Probleme lösen. Ein Handbuch*. Berlin: Jovis Verlag.

Sell, R., & Schimweg, R. (2002). *Probleme lösen. In komplexen Zusammenhängen denken. 6. Auflage*. Berlin: Springer.

Snowden, D. (2000, July). *Cynefin: A sense of time and place. An ecological approach to sense making and learning formal and informal communities*. Conference Proceedings of KMAC at the University of Aston.

Taleb, N. N. (2012). *Antifragile. Things that gain from disorder*. Great Britain: Clay's Ltd.

Ulwick, A. W. (2005). *What customers want. Using outcome-driven Innovation to create breakthrough products and services*. New York, NY: McGraw-Hill.

Weibel, B. (2014). *Simplicity—Die Kunst, die Komplexität zu reduzieren*. Zürich: Verlag Neue Zürcher Zeitung.

Zeichen, G. (2014). *Ingenieure an die Schalthebel. Mit den Fähigkeiten der "Komplexkönner" zu unternehmerischen Spitzenleistungen*. Wien: LINDE Verlag.

Zwicky, F. (1966). *Entdecken, Erfinden, Forschen im morphologischen Weltbild* (p. 51 ff). München/Zürich: Droemer Knaur.

TRIZ Evolutionary Approach: Main Points and Implementation

Victor D. Berdonosov and Elena V. Redkolis

Abstract There are a lot of evolution models, but the most interesting among them is the model of the TRIZ evolution. According to this model, artificial systems satisfy increasing demands of the society and evolve when they eliminate contradictions using TRIZ tools.

The TRIZ evolutionary approach can be applied in any sphere of human activity. Initial objects can be either engineering systems or knowledge systems of mathematics, programming, etc. In this case TRIZ evolutionary map is created for a considered sphere. Such a map is a tree of states (implementations) of the initial object which are connected by contradictions and TRIZ tools that eliminate these contradictions.

The TRIZ evolutionary approach provides an efficient tool for analyzing artificial system evolution and forecasting their development. Particularly, the TRIZ evolutionary approach allows defining contradictions eliminated and unsolved in a considered system. Thus if "forgotten" contradictions are eliminated, then system ideality will increase, respectively.

In the report there are examples of TRIZ evolutionary maps and forecasts of some artificial system development based on the TRIZ evolutionary approach.

Keywords TRIZ evolution • TRIZ evolutionary map • Artificial systems • Knowledge systems

1 Introduction

The theory of evolution is perceived subconsciously as evolution of biological species. Currently, there are a lot of theories and models of biological species evolution (Jukes and Cantor 1969; Kishino and Hasegawa 1989). Among them

V.D. Berdonosov (✉)
Komsomolsk-on-Amur State Technical University, Komsomolsk-on-Amur 681024, Russia
e-mail: berd1946@gmail.com

E.V. Redkolis
Komsomolsk-on-Amur State Technical University, Komsomolsk-on-Amur 681000, Russia
e-mail: redkolisev@gmail.com

L. Chechurin (ed.), *Research and Practice on the Theory of Inventive Problem Solving (TRIZ)*, DOI 10.1007/978-3-319-31782-3_6

there are modern evolutionary synthesis based on Darwin's and Mendel's (Mendel 1866) works, the model of molecular evolution that is known as the theory of neutrality, etc. However, artificial systems also evolve (Capra 1996; Shpakovsky 2006). It is of great interest to research evolution of these systems. Besides, now methods and approaches to the knowledge systematization in different knowledge fields receive considerable attention (Petrov 2012; Gurevich and Yashina 2012; Kapon 2015; Kumazawa 2014).

Unfortunately the theory of evolution is mostly evaluating but not forecasting issue. In our opinion the forecasting issue is the most important in researches of artificial systems. It means that if the evolution trend line of a system is known, then it is possible to forecast the following implementation of this system. TRIZ, research subject of which are technical systems, can also cover other artificial systems. Basing on TRIZ it is possible to create a model of evolution that is characterized by both evaluating and forecasting issues. The article describes the abovementioned model.

The model can be used only for artificial systems. An initial object exists in any evolution model, as for fractal model it is a pattern. Development of an artificial system is performed under moving forces in accordance with a set of rules.

2 Model of TRIZ Evolution

The TRIZ evolutionary model is ideologically associated with the fractal model though does not copy it. Mandelbrot (1983) was the fractal research founder. He described development (evolution) of fractal objects in the following way. There is an initial object ("a pattern") of free complexity. It can be either lines joined together or a multi-figure object. Then the pattern starts developing in accordance with "rules of construction." Each element of the pattern is replaced with its transformed copy. The transformation is a scaling, shortening up to the size of replacing element and rotating if it is required. This is the way how the first iteration of the fractal object creation is performed.

Following iterations are performed in the same way. Figure 1 shows the sequence of iterations: zero (the pattern), the third, the fifth, and the eighth. The iteration number is not limited. If a pattern is chosen and a considerable number of iterations are performed, then the fractal model will be similar to a real object. Example of the model is a fern.

Though ferns are simple plants, they have more difficult patterns and more complicated "rules of construction" which determine developing from iteration to another one. There are the following basic terms of the fractal model of evolution: pattern (initial object) is an object based on which the construction of a fractal object begins. The "rules of construction" are the rules according to which an iteration based on the previous one is performed. Moreover, there is one more term here—resources. For a fractal object, it is a space where the object is placed. For real fern resources are nutrients, space, and ultraviolet (sunlight).

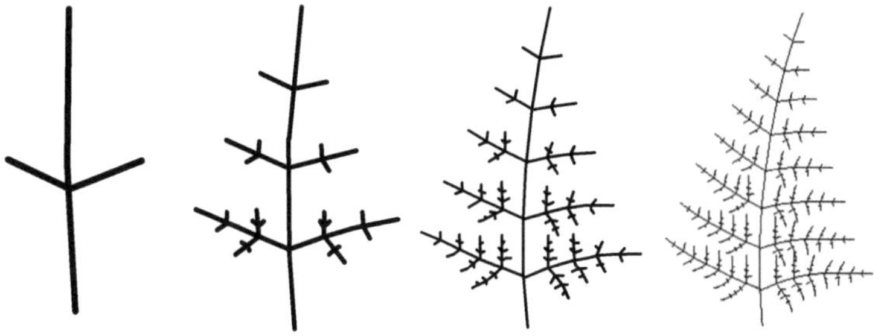

Fig. 1 A fractal model of fern (Berdonosov and Redkolis 2014b)

It should be noted that even for such simple plants as ferns, "patterns" and "rules of construction" are more complex than in the example given above. However, all plants are based on principle of self-similarity. Animals have more complex self-similarity. All necessary information about patterns is placed in animal genes (Capra 1996), and laws of nature determine "rules of construction." By the way, knowledge of these laws is the basic purpose of knowledge acquiring.

The evolution of the self-developing organisms can be represented in the following form: crystals, algae, corals, ferns, fishes, highest plants, birds, mammals, and human. Currently, a human is characterized by the highest level of complexity. It is supposed that human fractality should be performed not only on a physical level but on a spiritual level that is in consciousness. With the help of consciousness, a human explores the world and acquires a system of knowledge. Thus it is natural to suppose that the knowledge is also fractal. In fact, knowledge is the reflection of the world picture; thus, if the world is fractal, knowledge is fractal too.

The TRIZ evolutionary model uses the same set of terms as it is described above. A "pattern" is an initial artificial (technical) object which is a starting point for development of a family of objects.

It is necessary to clarify here what the word "technical" means. The Greek word "*techne*" was understood widely: from skills of craftsmen to proficiency in the high art. In the text the concept "technique" includes both a device (a machine) and a computer program that will control this device (machine), programming languages by virtue of which the devices have been created and developed (Blackburn 2014).

The next concept is "rules of construction" by virtue of which the initial pattern develops passing from iteration to another one. It is necessary to add some comments. At first the passage does not happen on impulse but under the action of contradictions revealed at the iteration between increasing requirements of the society and limited capacity of an object. The "rules of construction" of the TRIZ evolutionary models are TRIZ tools: 40 inventive principles, substance-field model, system of inventive standards, etc. (Altshuller et al. 2005; Zlotin 1999).

To show a family of technical objects, programming paradigms are chosen as an example (Berdonosov and Sycheva 2011). The initial object is coding. Coding is a system of machine language codes which is interpreted for the certain microchip. This is the first and the most elementary paradigm of programming.

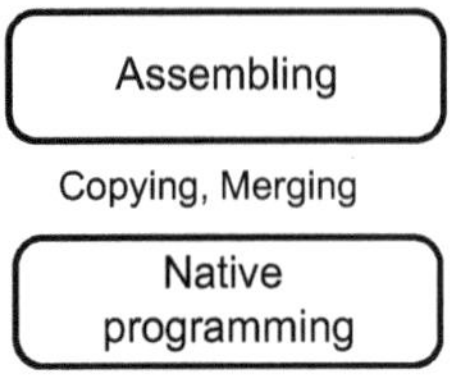

Fig. 2 The first iteration of programming paradigm evolution (Berdonosov and Redkolis 2014b)

Of course this paradigm has a lot of contradictions. Let's consider only one of them. Complexity of programming tasks increases constantly. The moment comes when a programmer cannot write and debug software which contains thousands of bits. Nowadays it is a very small program. For example, a program of calculating factorial function consists of not <1000 and half bits. Here the contradiction appears: if complexity of a task increases, then the time for programming increases unacceptably. To solve the contradiction, inventive principles of "copying" and "merging" can be used. Let's consider solutions that eliminate the contradiction. The first solution: using the principle of "copying," terms "cell address" and "argument" are replaced by the term "operand." The second solution: using the principle of "merging," homogeneous machine codes are merged into mnemonic commands.

In consequence of eliminating contradictions, a new resource appears. It is option to transform machine codes into mnemonic code. As a result, a new programming paradigm appears, that is, assembling (see Fig. 2).

TRIZ evolutionary model includes a "pattern" (an initial technical object) and the "rules of construction" which are TRIZ tools and resource limits.

It should be noted that TRIZ evolutionary model is proposed for structuring knowledge in subject area in terms of ideality increase. The ideality is a relation of the sum of parameters characterizing profits to the sum of the parameters characterizing expenses of the considered system. The simplification as an estimate of ideality is used in practice. This estimate is some specific parameter characterizing the quality of a system functioning. Thus, TRIZ evolution of a particular subject area can differ from evolution based on the historical retrospective or chronometric data.

In the context of programming paradigm evolution, parameters characterizing ideality are productivity, readability, and reliability of a programming language. In case of the numerical method evolution (which is discussed in Sect. 4), faithfulness, convergence, and number of mathematical actions can be considered as such parameters.

3 TRIZ Evolutionary Approach

Let's consider the TRIZ evolutionary approach which is characterized with the following statements:

The First Statement The evolution of a system starts from a base element (a system) (Shpakovsky 2006). This may be an incipient element or a prominent part, a stage of development of an incipient element. The examples of incipient elements are a wheel, a computer programming language, etc. (Fig. 3). The first wheel was probably a tree saw cut. The first programming language is defined as native programming language (Berdonosov 2012).

A system that heralded a famous stage of the wheel development was a spoke wheel. A system that heralded a famous stage of the modern programming language development was the first representative of object-oriented programming languages, which is Simula-67.

The Second Statement Moving forces of TRIZ evolution are contradictions between growing requirements of the society to a system and limited capacity of this system.

The owner of a wheel would like the wheel to work longer, but it was required to make the wheel rim thicker (Fig. 4). Wheels became heavy and horses drawing a cart were tired quicker.

The contradiction appeared: if durability of a wheel increases, then the weight of a wheel increases *unacceptably*. The same happens with object-oriented programming. A programmer writes more and more complex programs. Debugging complex programs, he spends much time to find causes of errors. Thus, the contradiction

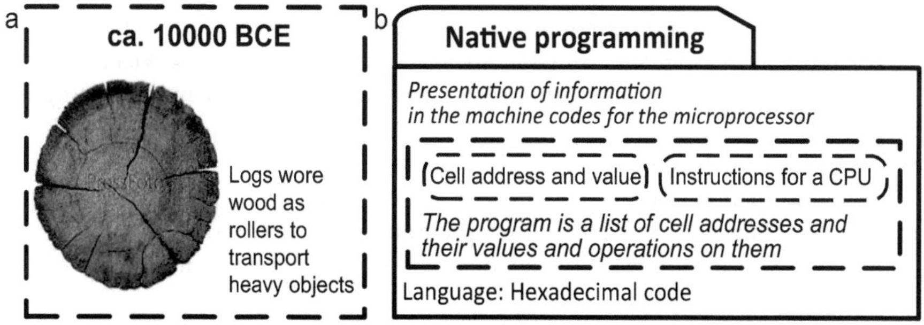

Fig. 3 (**a**) A tree saw cut; (**b**) native programming (Berdonosov and Redkolis 2014a)

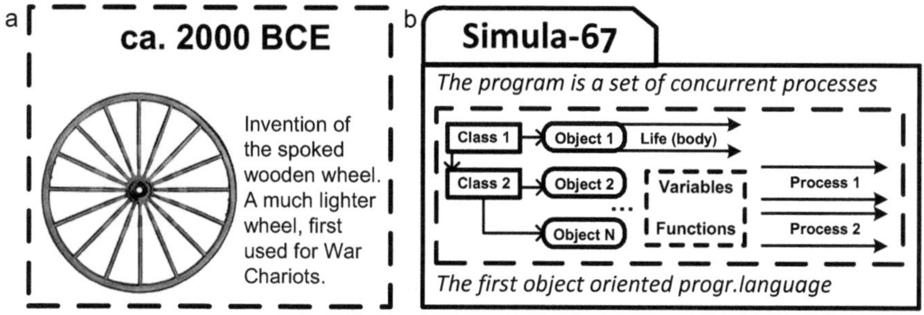

Fig. 4 (**a**) A spoke wheel; (**b**) the structure of Simula-67 programming language (Berdonosov and Redkolis 2014a)

appears: if complexity of a program increases, then the time for debugging increases *unacceptably*.

De facto, there is not one but a variety of contradictions (Berdonosov 2012). In case of object-oriented programming languages (OOPL), more contradictions appear: if the size of a program increases, then reliability of the program decreases *unacceptably*; if the quantity of hardware platforms increases, then working efficiency of a program decreases *unacceptably*.

The Third Statement A system passes to the following stage of evolution when contradictions are eliminated. It is always possible to reveal TRIZ tools which allowed eliminating contradictions (40 inventive principles, substance-field model, system of inventive standards, etc.) (Gadd and Goddard 2011; Bukhman 2012; Fey and Rivin 2005; Livotov and Petrov 2013).

In case of a wheel, inventive principles of "local quality" and "intermediary" were used. The wooden wheel is covered with a metal rim, the weight of the wheel decreases, and durability of the wheel increases. In case of the OOPL, part of contradictions was solved in Smalltalk language using the law of transition to supersystem: software development environment was created. It had a user interface and provided debugging tools. By the principle of "intermediary," the sequence of program compilation was changed: programs were coded to intermediate form by byte-codes and interpreted into a machine language code during the run-time. It allowed initializing them on different hardware platforms (Berdonosov 2012). In the C++ language using the inventive principle of "self-service," the tool of handling exceptions was introduced. This tool is used to monitor program behavior and catch errors (Berdonosov 2012). A tool "design by contract" was performed in the Eiffel programming language using the inventive principle of "preliminary action." This tool allowed assigning different types of conditions (contracts) which are checked during a program run-time (Berdonosov 2012).

Initial part of the OOPL TRIZ evolutionary map is presented in Fig. 5.

The Fourth Statement All stages of a base element (system) evolution are visualized with the TRIZ evolutionary map. Such a map is a tree where the base element is a root and stages (iterations) of evolution are branches (see Fig. 6).

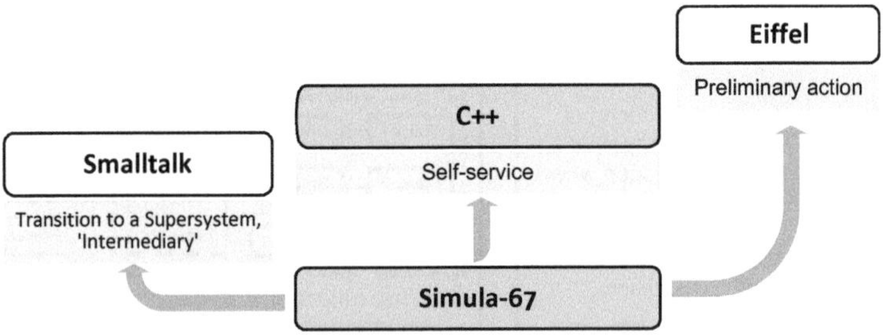

Fig. 5 First iteration of object-oriented program language evolution (Berdonosov and Redkolis 2014b)

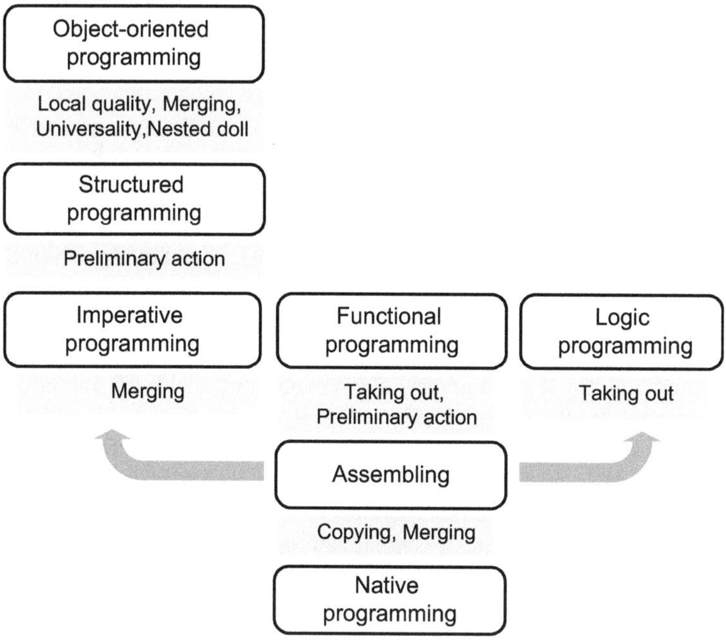

Fig. 6 Tree-type evolution of programming paradigm (Berdonosov and Redkolis 2014b; Berdonosov and Sycheva 2011)

It should be noted that in reality fragments of the tree can grow linearly if the contradiction is eliminated by one tool or fork out if different contradictions are eliminated by different tools. In the last case, the line evolution transforms into a tree-type one (Fig. 6). As a result, the TRIZ evolutionary approach to artificial system analysis allows not only systematizing knowledge about its evolution but also inventing (developing) new implementations.

Thus, there are following advantages of the TRIZ evolutionary approach:

1. It allows systematizing knowledge about artificial system evolution in past, present, and future basing on TRIZ tools in order to forecast further development of artificial systems.
2. It allows inventing (developing) new implementations of artificial systems.
3. It provides an integrated base for knowledge systematization in any field either practical (e.g., TRIZ evolution of a car or a plane) or theoretical one (e.g., TRIZ evolution of numerical methods, TRIZ evolution of programming tools, etc.).
4. It allows sorting system implementations according to the rate of ideality increase at the stage of analysis and research of new system implementations.

4 Examples of the TRIZ Evolutionary Approach Implementation

Let's consider full implementation of the TRIZ evolutionary approach by the example of numerical methods (Korn and Korn 1961). Numerical methods are defined as methods of approximate solution of typical mathematical problems, which perform finite quantity of elementary number operations. These methods are various (Altshuller et al. 2005; Mandelbrot 1983; Shpakovsky 2006; Danilina 1976): linear algebraic equation system solution and equations; nonlinear algebraic equation system solution, numerical integration, and differentiation; solution of Cauchy problem for ordinary differential equation; etc. In addition there is a list of specific methods for almost every abovementioned field of application (see (Berdonosov and Redkolis 2014b), Figs. 7, 8, 9, 10, 11, 12, 13, 14, and 15).

Such a significant number of poorly systematized knowledge (methods) does not allow studying all numerical methods sufficiently in order to choose the most suitable one and for education process too.

There are many mathematics systematizations and numerical method systematizations; it complicates choosing a suitable method. So there are many approaches to organize and transfer mathematical knowledge, for example (Wilkerson-Jerde Michelle and Wilensky Uri 2011). Analysis of numerical methods classification showed that numerical methods evolve both during development of description models of real physical objects and during development of problem-solving methods represented by the corresponding models. Thus it is useful to consider TRIZ evolutionary maps of models and numerical methods, relative to each model.

The evolution of models is associated with increasing mathematical model adequacy to their real physical prototype. For example, behavior of different nature macro-model systems is described by linear and nonlinear algebraic equation systems; behavior of micro-models is described by differential equation systems; and behavior of distributed system micro-models is described by differential equations in the form of partial derivatives. The evolution process of numerical methods is concerned with ideality increase of existing model realization. For example, first direct numerical methods were used for linear algebraic equation solution, then iterative one-step methods and iterative multistep methods followed, etc. Whereupon the ideality criterion consists of accuracy, convergence, number of arithmetic operations, etc.

Let's consider the development of real physical object description models, i.e., the development of mathematical models which describe objects of the real world more and more adequately.

The first models were linear equation systems. But scientists found out that if an acceptable region of variants that are included in the equation system is wide enough, then test data and estimated data will considerably differ. This situation appears because the world is not linear in principle and linearization can be performed in a small range. To bring estimated data more in line with test ones, a wide acceptable region was divided into small ranges, and parameters of linear

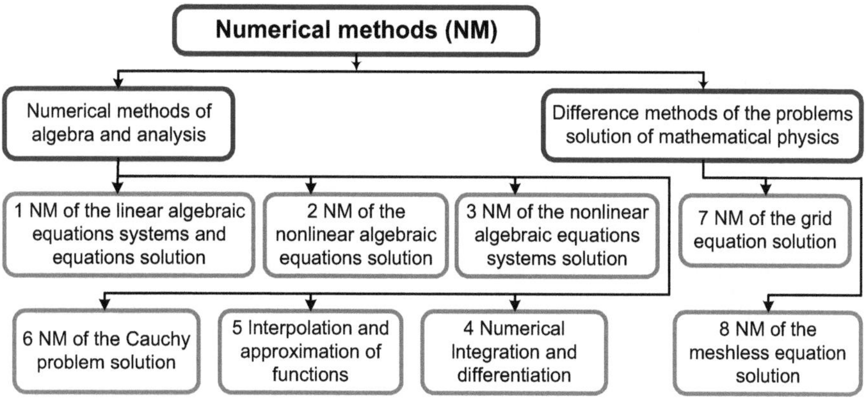

Fig. 7 Numerical method structure

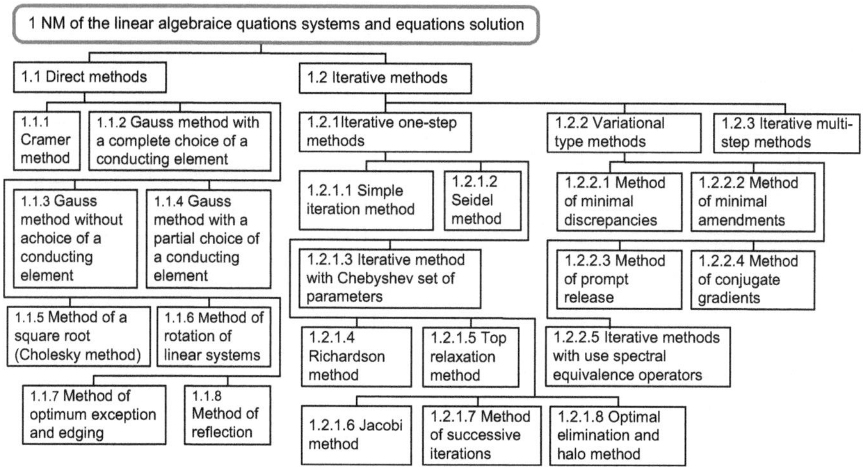

Fig. 8 Numerical methods of linear algebraic equation systems and equation solution

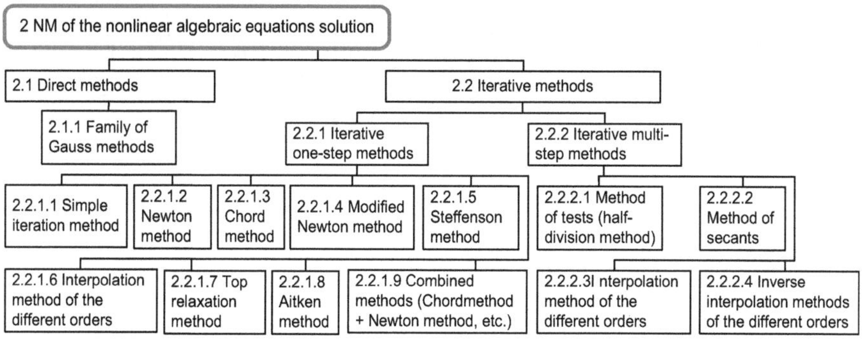

Fig. 9 Numerical methods of nonlinear algebraic equation solution

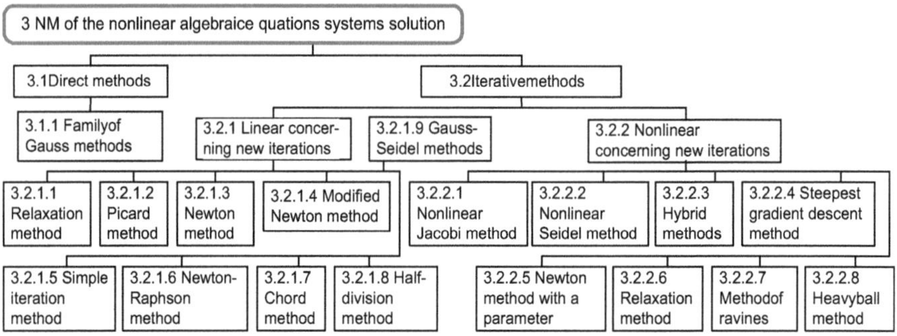

Fig. 10 Numerical methods of nonlinear algebraic equation system solution

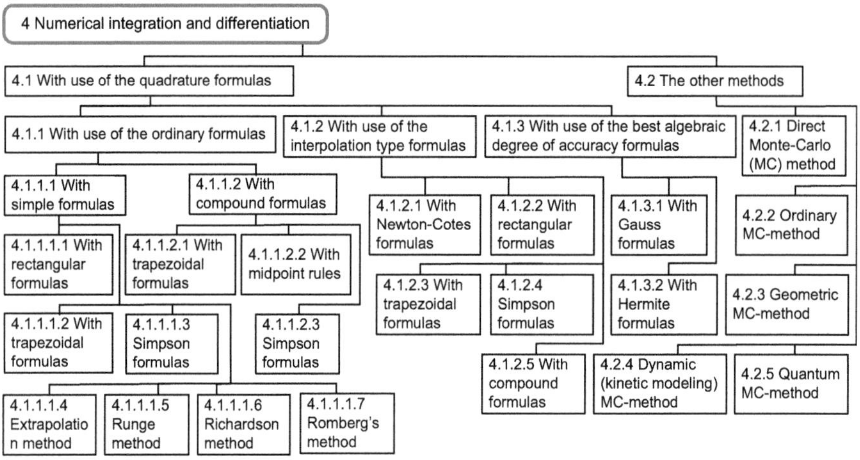

Fig. 11 Methods of numerical integration and differentiation

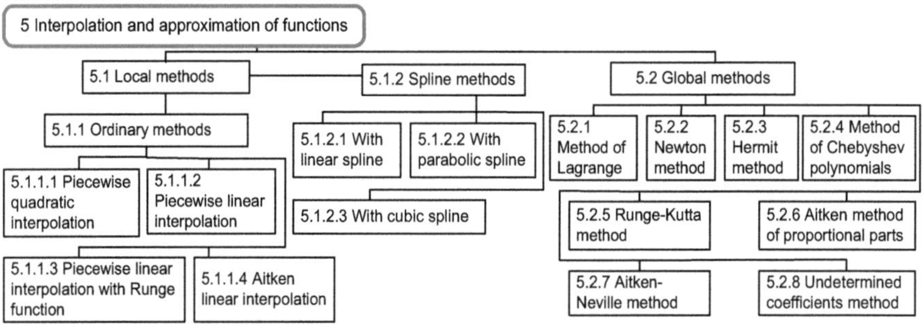

Fig. 12 Numerical methods of interpolation and approximation of functions

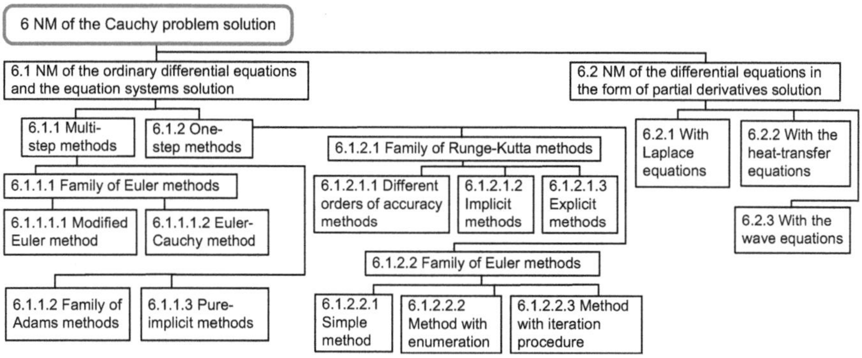

Fig. 13 Numerical methods of the Cauchy problem solution

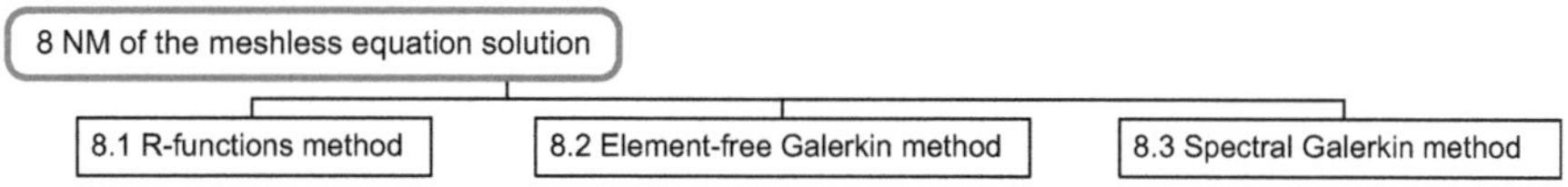

Fig. 14 Numerical methods of the meshless equation solution

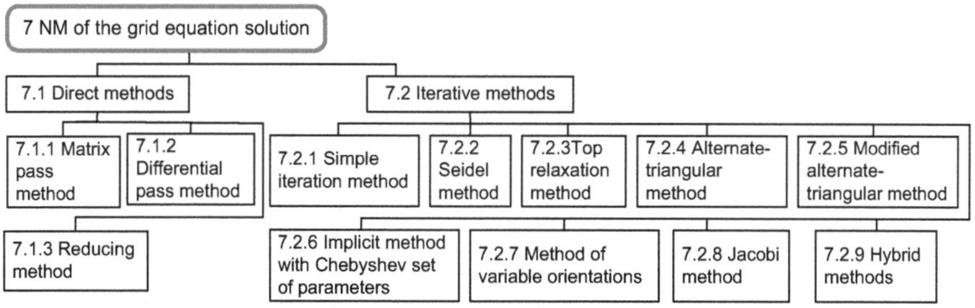

Fig. 15 Numerical methods of the grid equation solution

equitation systems were defined separately for each range. It led to the great volume of calculations.

Thus, there is a contradiction: increase of closeness of agreement of linear equation system solution to test data led to *unacceptable* increase of calculation amount. To solve this contradiction, the principle of "another dimension" (Altshuller et al. 2005) was used, where "another dimension" means transition to the category of nonlinear functions (equations).

The transition to nonlinear equations allowed describing technical object functioning more or less adequately but only in one field subsystem. For example, the welding process had been described only in electrical subsystem. However, for any real technical object, there are processes referring to different subsystems. In the welding process, electrical subsystem is only initial one, and then heat subsystem appears and then deformation, hydraulic, and other subsystems. While solving nonlinear equations for different subsystems that are not connected with each other, significant mistakes appear. There is a contradiction, which can be eliminated by the principle of "merging" (Altshuller et al. 2005). A technical object can be described with the system of nonlinear equations, etc.

Eliminating contradictions that appear during development of numerical methods, the line of evolution of description models of real physical objects is described (see Fig. 4.10). Similarly, the evolution of the individual methods for solving tasks is considered.

Let's consider the first model, which is the method of solution of linear algebraic equations. Let it is required to increase method's rate of convergence (to decrease number of final arithmetic calculations). The rate of convergence is limited by the number of arithmetic calculations for direct and counter operations (time for performing direct and counter operations). Time for operations performing depends directly on a type (a structure) and a degree of a matrix: solution will be found faster if degree is less. The case with structure of a matrix is similar, because solution will be found faster if structure is simpler (Danilina 1976).

Therefore, to decrease the number of final arithmetic calculations, it is necessary to make a mathematical model of the real world that contains a matrix of a medium degree (<100) or, and that is better, of a small degree which angle minors shall be nonzero. But such mathematical model does not fully describe the behavior of system macro-models (Fig. 16).

This mathematical model does not completely describe the behavior of macro-models of different nature systems. There is a contradiction: with increasing rate of the Gauss method convergence, the number of real object macro-models decreases *unacceptably*. To eliminate this contradiction, it is proposed to use the principle of

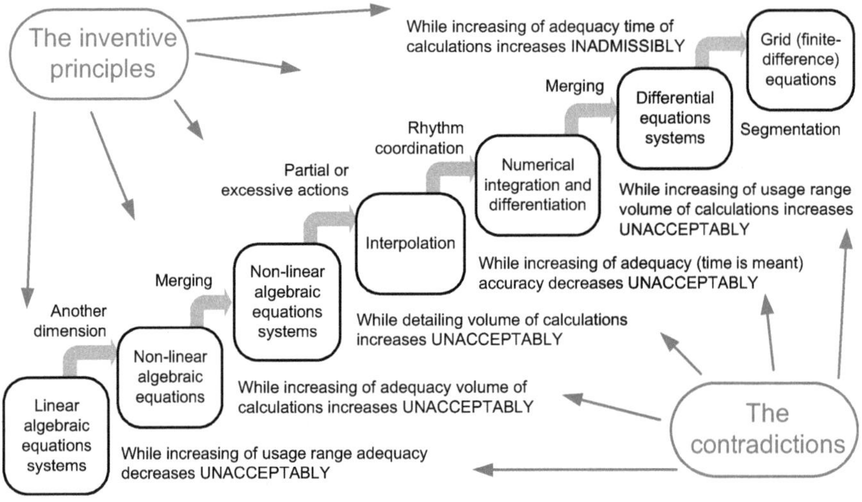

Fig. 16 Evolution of the object description (Berdonosov and Redkolis 2014b)

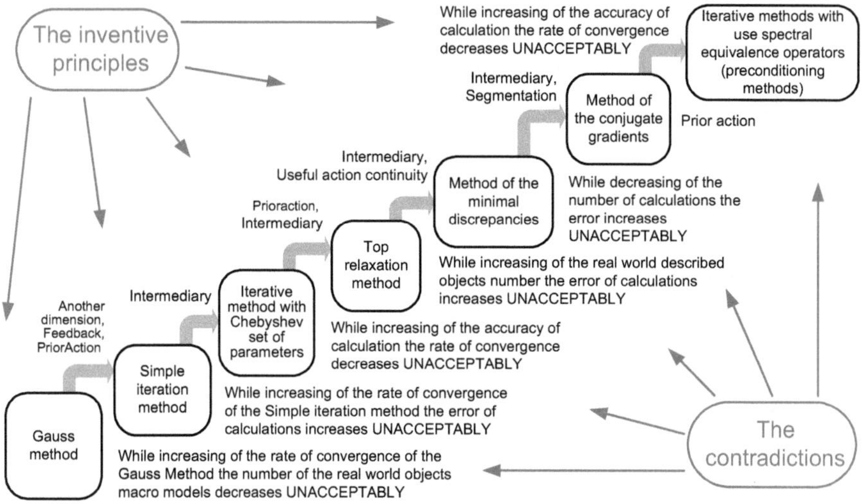

Fig. 17 Evolution of methods of linear algebraic equation system solution (Berdonosov and Redkolis 2014b)

"feedback" (Altshuller et al. 2005). Numerical simple iteration method was used. Analyzing other solution methods of linear algebraic equation systems, linear TRIZ evolutionary map can be constructed (see Fig. 17).

After reviewing all the basic lines of method development, it is possible to construct TRIZ evolutionary map of numerical methods (see Fig. 18).

Comparing the created TRIZ evolutionary map (Fig. 18) and the initial classification, it is noted that:

1. Simple and demonstrative presentation of the numerical method structure was created that can be used in education if there is limited time to study a significant volume of material.
2. Development of numerical methods as other artificial systems (e.g., programming paradigms) is subordinated to clear logic of ideality increase.
3. Quantity of "basic" numerical methods that are required to learn for solving different problems were reduced due to excluding methods that have similar ideality.

TRIZ evolutionary map (see Fig. 18) considers only evolution of computational methods of mathematical tasks solving in terms of numerical methods and does not consider all of mathematical methods known for today.

It should be noted that the use of TRIZ evolution map is suitable for a particular subject area. Such area is a group of systems designed to solve a particular class of problems (in mathematics, it is computational mathematics, mathematical statistics, mathematical logic and chaos theory, etc.). For these groups, it is necessary to build the different TRIZ evolutionary maps, carefully defining operating parameters characterizing systems of each group.

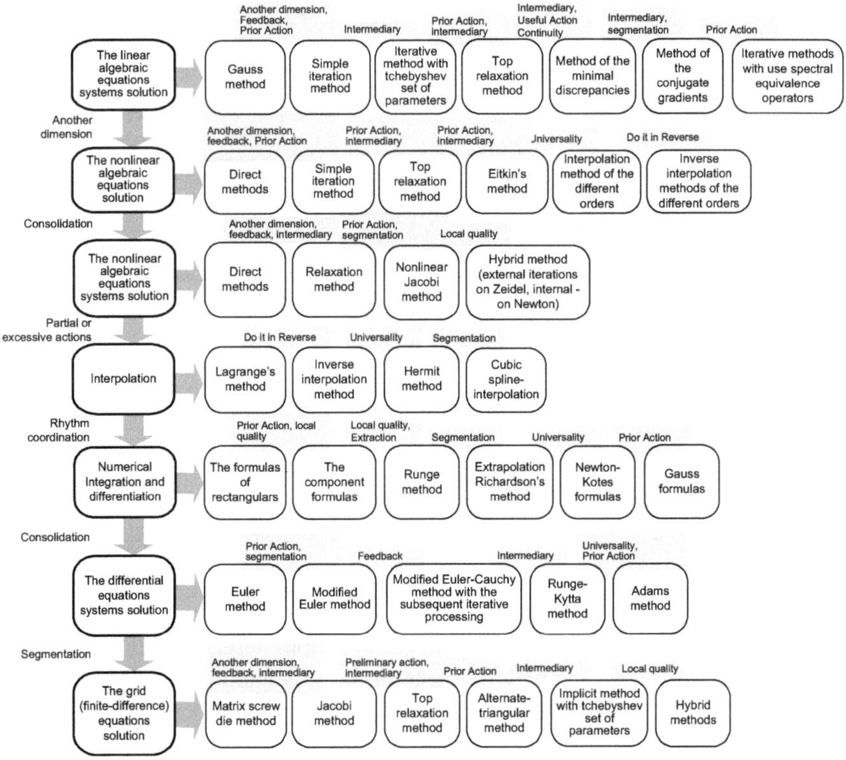

Fig. 18 TRIZ evolutionary map of numerical methods (Berdonosov and Redkolis 2014b)

By this means, TRIZ evolutionary approach can be successfully applied to knowledge systematization of objects and processes not only in technical but also in any other fundamentals. Such example of the TRIZ evolutionary approach application in nontechnical scientific knowledge is budgeting. It is a process of the organization management, which includes coordinated planning and control of financial and economic state with distribution of responsibility for results of operation. Evolution of budgeting paradigms in accordance with the TRIZ evolutionary approach is shown in Fig. 19.

Our team has been developing TRIZ evolutionary maps for systems from different subject fields to confirm this thesis for several years.

So, the first map was made for addressing systems of a dynamic access memory. It was called one-dimensional and not TRIZ evolutionary yet. Then, the map for numerical methods was developed (see Fig. 18). This map was constructed as a composition of several linear maps. One line on this map was the development of methods for describing real objects representing the equation or system of equations. Other one-dimensional maps, representing development of methods for solving equations (or equations systems), were attached to each element of this line. Then maps were constructed for a number of systems related to the software:

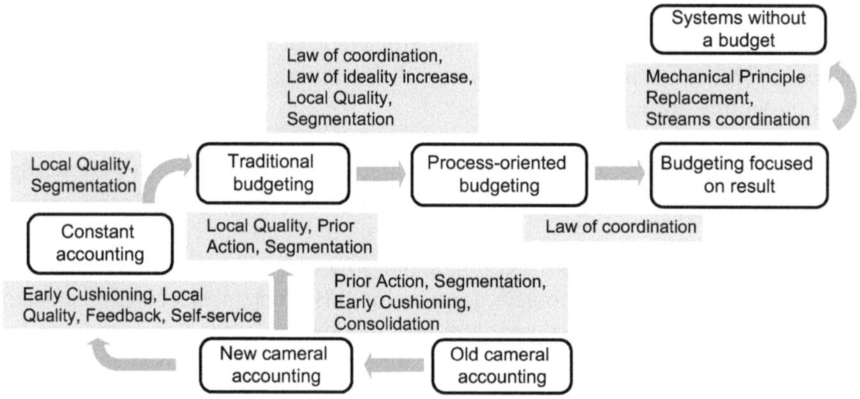

Fig. 19 TRIZ evolutionary map of budgeting paradigm

computer-aided software engineering systems, programming paradigms, object-oriented programming languages, mechanisms of object-oriented programming languages, etc. As the milestone, it was concluded that the most common structure of a map is a tree (linear and composition of linear structures are special cases of a tree structure).

In addition, for more evident and compact representation of objects, which are map elements, it was offered to use the "frames." Frame is a structure that contains the description of the object, process, event, phenomenon, or situation in the form of some pattern, while a described system exists within a certain paradigm of scientific knowledge.

Finally, considering these statements, a map was constructed for a system from a nontechnical subject area. It was TRIZ evolutionary map of budgeting paradigm (see Fig. 19).

The work dedicated to constructing TRIZ evolution maps continues now. The following maps are in process: statistical methods (mathematics), manufacturing execution systems (management), and expert systems (software). Thus, it can be said with confidence that even such nontraditional subject areas for TRIZ as mathematics, programming, economics, and management can be systematized by means of TRIZ evolutionary approach.

5 Discussion and Conclusions

TRIZ is an excellent tool for systematization of a wide range of artificial systems according to researches (Berdonosov and Redkolis 2011, 2014a, b; Berdonosov and Sycheva 2011; Berdonosov 2012, etc.). TRIZ evolutionary map is as a result of such systematization. The basic TRIZ trend lines are presented in a map: increasing ideality through contradiction elimination, roll-in–roll-out, etc.

If TRIZ is used as a foundation for analysis of different artificial systems (from technical to knowledge systems), the efficiency of educational process will increase. The educational process is aimed at learning main types and objective laws of definite artificial systems.

Totally, it can be concluded that usage of the TRIZ evolutionary approach allows intensifying the study of knowledge areas by systematization of knowledge not only at the level of contradiction resolution but also at the level of contradiction formulation.

The way of thinking "from a contradiction to a contradiction" allows systematizing information about a problem, formalizing this problem, selecting parameters which are improved or worsened, and formulating it clearly that speeds up the search of solution. Sometimes the correct question is already the best answer.

However, it should be understood that the active use of the TRIZ evolutionary approach to the knowledge systematization in the educational process requires the preparatory activities. In particular, it is necessary to develop some reference models of TRIZ evolution for different subject areas. Obviously, the evolution mechanisms ("rules of construction") for the systems belonging to the natural or technical studies may be different from the evolution mechanisms of the humanitarian or social systems. The existence of such differences, first of all, is explained by the different definition of objects and subjects in studies of different classes.

In addition to education, another perspective field of TRIZ evolutionary approach application is its use for systematization of information about a subject area in order to forecast the demand for the creation of a new system and control the evolution of a system group. Currently the authors unify and generalize the terminology and develop ways to apply TRIZ evolutionary approach in order to predict the development of artificial systems, considering types of contradictions. Also, if the TRIZ evolutionary approach is used, engineering quality increases at stages of analysis and proposals (researches) of new implementations of technical systems.

References

Altshuller, G., Clarke, D. W., & Fedoseev, U. (2005, April). *40 principles: TRIZ keys to innovation* (Extended ed.). Paperback. MA: Technical Innovation Center. ISBN-13:978–0964074057, ISBN-10:0964074052.

Berdonosov, V. D. (Ed.). (2012). *TRIZ evolution of the object-oriented programming languages. Proceedings of the ETRIA TRIZ future conference*. Lisbon: University Nova de Lisboa.

Berdonosov, V. D., & Redkolis, E. V. (2011). TRIZ fractality of mathematics. *Procedia Engineering, 09*, 461–472. doi:10.1016/j.proeng.2011.03.134.

Berdonosov, V. D., & Redkolis, E. V. (2014). *TRIZ evolutionary approach: didactics*. ETRIA World Conference TRIZ Future 2014. Global Innovation Convention.

Berdonosov, V. D., & Redkolis, E. V. (2014). *TRIZ evolutionary approach: Main points and implementation*. The 10th International MATRIZ Conference, TRIZfest-2014: Theories and Applications.

Berdonosov, V. D., & Sycheva, T. (2011). *TRIZ evolution of programming system*. ETRIA World Conference TRIZ Future 2011. Dublin: Institute of Technology Tallaght.

Blackburn, S. (2014). *The Oxford dictionary of philosophy* (2nd rev. ed.). Oxford University Press. The Print ISBN-13:9780199541430. Accessed 2008, from http://www.oxforddictionaries.com/definition/english/technical

Bukhman, I. (2012). *TRIZ technology for innovation*. P.049 "3.3.2 The law of nonuniform evolution of subsystems (system components). Creating System Contradictions.

Capra, F. (1996). *Web of life: A new scientific understanding of living systems*. New York, NY: Anchor Books.

Danilina, N. (Ed.). (1976). *Numerical methods. Manual for technical schools*. Moscow: The Higher School.

Fey, V., & Rivin, E. (2005). *Innovation on demand: New product development using TRIZ*. ISBN-13:978–0521826204, ISBN-10:0521826209.

Gadd, K., & Goddard, C. (2011). *TRIZ for engineers: Enabling inventive problem solving* (p. 504). UK: Wiley. ISBN ISBN:978-0-470-74188-7.

Gurevich, I. B., & Yashina, V. V. (2012). Descriptive approach to image analysis: Image formalization space. *Pattern Recognition and Image Analysis, 22*(4), 495–518. doi:10.1134/S1054661812040050.

Jukes, T. H., & Cantor, C. R. (1969). *Evolution of protein molecules* (pp. 21–132). New York, NY: Academic.

Kapon, S. (Ed.). (2015). Perceiving permutations as distinct outcomes: The accommodation of a complex knowledge system. *Educational Studies in Mathematics, 88*(1), 43–64. doi:10.1007/s10649-014-9570-2.

Kishino, H., & Hasegawa, M. (1989). Evaluation of the maximum likelihood estimate of the evolutionary tree topologies from DNA sequence data, and the branching order in hominoidea. *Journal of Molecular Evolution, 29*(2), 170–179.

Korn, G., & Korn, T. (1961). *Mathematical handbook for scientists and engineers*. New York, NY, Toronto, London: McGraw-Hill.

Kumazawa, T. (Ed.). (2014). Initial design process of the sustainability science ontology for knowledge-sharing to support co-deliberation. *Sustainability Science, 9*(2), 173–192. doi:10.1007/s11625-013-0202-z.

Livotov, P., & Petrov, V. (2013). *TRIZ innovation technology. Product development and problem solving. Handbook* (p. 280).

Mandelbrot, B. (1983). *The fractal geometry of nature*. New York, NY: Freeman.

Mendel, G. (1866). *Versuche über Pflanzen-Hybriden* (pp. 3–47). Verhandlungen des naturforschenden Vereins in Brünn, t. IV.

Petrov, T. G. (2012). Graphic representation of the evolutionary processes of the compositions of multicomponent objects of any nature. *Automatic Documentation and Mathematical Linguistics, 46*(2), 79–93. doi:10.3103/S0005105512020045.

Shpakovsky, N. (2006). *Trees of evolution. The analysis of engineering information and generation of new ideas*. Moscow: TRIZ profi.

Wilkerson-Jerde Michelle, H., & Wilensky Uri, J. (2011). How do mathematicians learn math? Resources and acts for constructing and understanding mathematics. *Educational Studies in Mathematics, 78*(1), 21–43. doi:10.1007/s10649-011-9306-5.

Zlotin, B., Zusman, A., Altshuller, G., & Philatov, V. (1999). *Tools of classical TRIZ* (p. 266). Ideation International Inc. ISBN 1928747027. http://www.ideationtriz.com/publication.asp#Tools_of_Classical_TRIZ

Contradiction-Centred Identification of Search Fields and Development Directions

Verena Pfeuffer and Bruno Scherb

Abstract Anticipating changes in the environment of a company, aligning its strategy for the future to these changes and identifying new business fields are the tasks of the innovation management function within a company. In order to achieve structured and systematic action in this early phase of the innovation process, a TRIZ-based procedural structure was developed for deriving search fields and development directions.

Due to the different levels of detail of the search field and development direction, two separate structures were created, in which extrapolative and retropolative methods were combined in order to generate a holistic vision.

For the procedural structure for derivation of search fields, an extrapolative procedure is combined with a retropolative approach. The extrapolation contains a combination of TRIZ and foresight methods, which assist in deriving requirements. The interactions between these requirements are then analysed, and key contradictions are identified from this analysis. Overcoming the key contradictions between individual requirements is embodied in future-oriented search fields. The extrapolative approach is combined with retropolative methods such as scenario technique and/or visioning, in order to generate visions of the future and use these to deduce strategies for the future.

In addition to the procedural structure for deriving search fields, a further structure was defined for deriving development directions. It is fundamentally possible to derive development directions starting from existing search fields or from the solution space with an already existing product. If one starts with a search field, the procedural structure prepared for search fields can be adopted. If one starts in a specific solution space, the procedure begins with a history and product analysis, uses development front analysis to show the current state of development and continues with S-curve analysis, application of the development laws and trend screening.

V. Pfeuffer (✉) • B. Scherb
Schaeffler Technologies AG & Co. KG, Herzogenaurach, Germany
e-mail: pfeufvre@schaeffler.com

L. Chechurin (ed.), *Research and Practice on the Theory of Inventive Problem Solving (TRIZ)*, DOI 10.1007/978-3-319-31782-3_7

1 Classification in the Innovation Process and Relevance of the Subject

The innovation process as described in the literature is divided into the fuzzy front end, new product development and commercialisation phases (Koen et al. 2001). In a variation on this principle, Schaeffler regards the fuzzy front end as the Schaeffler innovation process, which is the responsibility of corporate innovation management. The fuzzy front end comprises the phases "identification, generation and evaluation". In the identification phase, the search fields and development directions are identified first, from which the problem issues are derived. In the generation phase, ideas are then developed to address these problem issues. In the third phase, the ideas are then evaluated. The best ideas are implemented in the advance development and product development process. In this case, the funnel form in Fig. 1 reflects the search space. This comprises the entire environment of the company, including trends, market situation, strategy of competitors and in-house competences. On the basis of these individual methods and the widely varying external and internal influences and pieces of information, search fields and development directions are derived.

A systematic procedural structure is to be prepared here, which can then be used to arrive at a structured analysis of the company's environment and, based on the information gained, to then derive search fields.

The requirement for action is the result of increasing globalisation. This leads to increasing market pressure with an intensification of international competition. Many market sectors are experiencing a paradigm shift. While, in the past, the

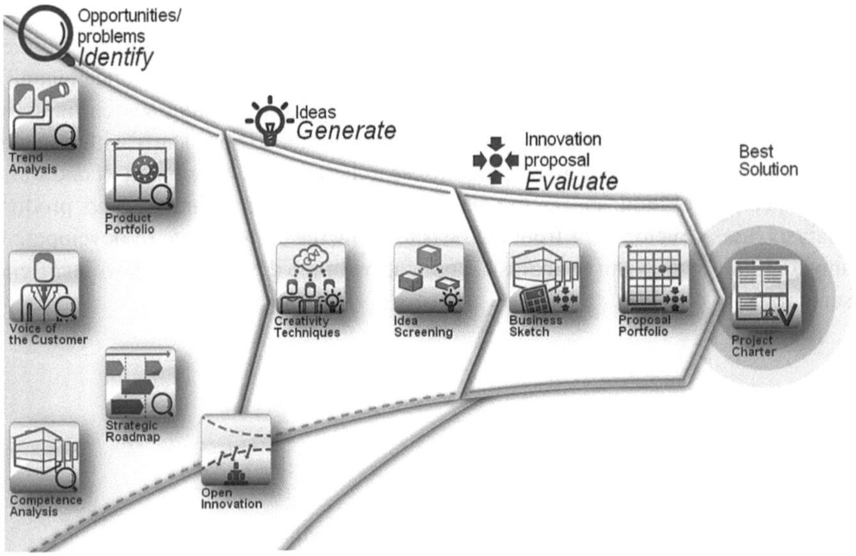

Fig. 1 Innovation process of Schaeffler AG & Co. KG (2015)

innovation cycles were long, plannable and in some cases even predictable, the picture is now of radical growth in the pace of innovation. New competitors are entering the market—in some cases from quite different sectors and new technologies are making it possible to achieve completely new solutions to problems. In a dynamic environment of this nature, the classic approaches applied in innovation management are too slow to keep pace with the changes. The strategy for survival in markets that are ceaselessly becoming more dynamic lies in increasing both the strength and speed of innovation.

These challenges can be confronted by paying particular attention to the identification phase of the fuzzy front end. However, only rudimentary systematic structures have yet been established in this respect. At the implementation level, the so-called back end, clearly structured processes for advance development and product development are applied in order to create milestones, activities, rules and project plans that will proceed through a stage-gate process. In the fuzzy front end of identification, searching for a direction and deciding on a direction, only individual methods are present, while a systematic structure in processing for deriving search fields and development directions is only available in the form of approaches.

2 Fundamental Definitions

For the development of a procedural structure, it is necessary to prepare a definition of the term "search field" as a basis for work and communication. Starting from this definition, other essential terms such as development direction, development forecast, development line and development front are defined and delimited.

The search field is defined as that element in the environment of a company that addresses trends covering usually more than one market sector and/or is characterised by new technologies. Search fields are the result of requirements that have not been adequately met, problems that have not been solved and/or on the basis of new possibilities. A search field describes the structural framework with defined boundaries.

Other criteria for a search field are:

- Significant global relevance
- Neutral in terms of function and solution
- A relevant time window with a fluid beginning and end
- A subject area of a consistent nature
- Conceptual focus on the future
- Several development directions that often result within one search field

In the development of a company's strategy for the future, the search fields relevant to the future must be identified and described. These are oriented to the company's own competences.

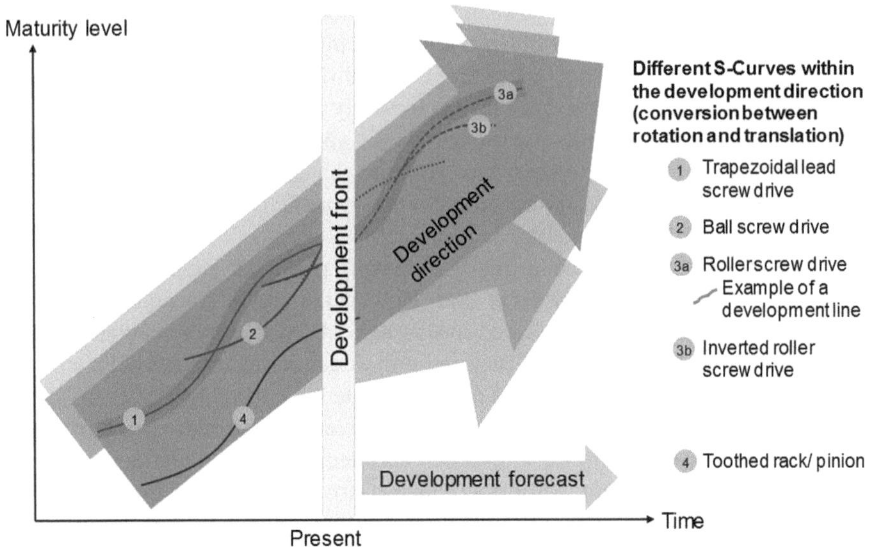

Fig. 2 Delimitation of the terms development line/forecast/front/direction

The term development direction is used to describe a key development issue within a search field that specifies the implementation of a function or places the focus on certain components. The development direction with the greatest potential is the extrapolative alignment of the current development front in the direction of an ideal system.

The delimitation of the terms development line, development forecast and development front of a development direction is shown in Fig. 2. The development directions are portrayed by means of arrows. Each development direction often contains more than one development line, which results in turn from the combination of individual S curves. The development front reflects the current state of technology, while the development forecast shows the probable continued development in the future.

For better understanding, an example from actuator technology can be used.

In order to facilitate conversion of rotational into translational motion, trapezoidal lead screw drives (1) are used first. In order to reduce their friction, ball screw drives (2) are developed later. Due to the replacement of sliding friction by rolling friction, friction is significantly reduced, which increases the efficiency level. In order to further increase the performance density and prevent slippage, ball screw drives are taken as a starting point for the development of roller screw drives (3a) and their subordinate manifestation, the inverted roller screw drive (3b). In parallel with the screw drives, rotation can be converted into translation with the aid of a combination of a pinion and toothed rack (4). Since the function of screw drives and the pinion/toothed rack combination is the same, all these are compiled in one development direction. The flatter arrows indicate that one development direction may be subdivided on the basis of different objectives. Parallel development directions in the search field can include microactuators or electric motors, which are indicated in the diagram by the two parallel arrows. While the S curves have been modelled for all screw drives and the pinion/toothed rack combination, only the development line of the roller screw drive is shown by the thick strip for the purposes of better transparency.

Development lines describe the development of technical systems along an S curve. In this process, the systems pass through the phases of development, growth, continuation, maturity and degradation. Through the development of a new system or a new generation, the transition to a new S curve can be created. The development line follows the jump to the new S curve and then travels along it. If a technical system gives rise to more than one new system, the development line is divided as shown in Fig. 2 for the S curves of roller screw drive and inverted roller screw drive (WOIS Institut 2008).

The development front describes the current state of technology and highlights variants for implementation within the same market sector or in areas outside the market sector. Due to the morphological procedure in the search for the different versions of functional implementation and the abstraction of individual subfunctions to basic functions, alternative concepts can be searched in a targeted manner. The development front is set by the current barriers to development that cannot yet be overcome with the current level of knowledge, the available technologies and the existing materials (Linde et al. 2008).

Through extrapolation with the aid of laws, the development forecast gives a forecast of the development directions with significant potential for the whole system and its subsystems. First, the development front is analysed in order to search the possibilities for expansion of their performance limits. The current barriers to development can be overcome in this way. The discrepancy in the development status of individual subsystems is also reflected in the development forecast, since this forecasts larger development steps for less developed systems (Linde et al. 2008).

3 Procedural Structure for Search Fields

Figure 3 shows in schematic form the concept underlying the procedural structure prepared.

The search fields to be derived are located at the centre of the diagram and are characterised by requirements and trends. One requirement, for example, is the wish for mobility. On the other hand, there is a trend towards increasingly stringent environmental rules and the increasing awareness of environmental protection. A search field reconciles different requirements and supplies solutions, for example, under the heading "Green Mobility". Within a search field, various development directions are generated that contain the solution spaces. The shape—similar to that of an hourglass—represents the process of narrowing down from the requirements and trends to the search fields and the subsequent opening out into the space for

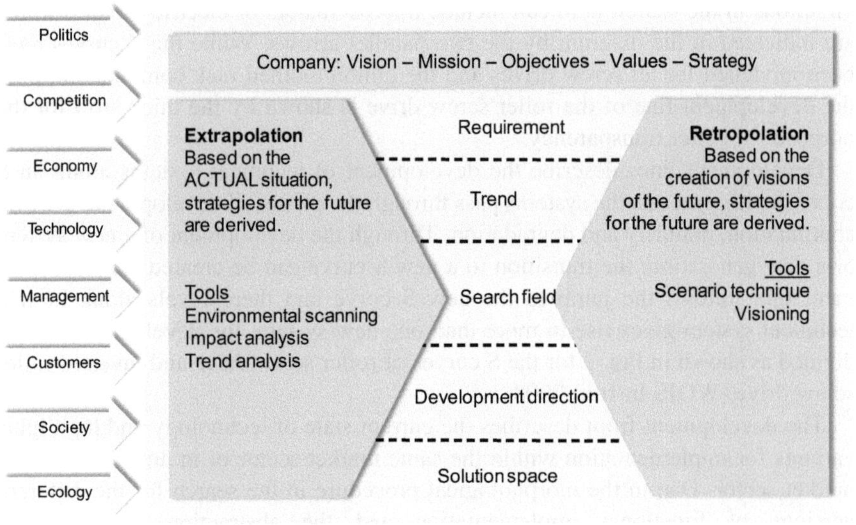

Fig. 3 Classification of search fields

development directions and solution spaces. For the derivation of strategies for the future with the aid of search fields, extrapolative and retropolative methods are combined with each other. Extrapolation contains an analysis of the current situation (actual state analysis) and in cases also the past (history analysis) and uses the evolution laws to present these as a forecast of developments extending into the future. With the aid of retropolation, desired visions of the future are created, from which strategic subject areas are derived in order to generate a long-term perspective. The extrapolative and retropolative images are overlaid, thus giving a comprehensive vision of the future. All the derived search fields must not contradict the vision and mission, objectives, values and strategies of the company. The entire system and the procedural structure are characterised by external influences, such as the economic situation or social structures. For the derivation of search fields, all these influences must be taken into consideration, identified and analysed.

In the extrapolative approach, different individual methods such as environmental scanning and impact analysis are combined with the central idea taken from TRIZ, the formulation of contradictions.

The retropolation comprises the scenario technique shown on the middle branch and the visioning shown on the right. Both are based on the assumption of the influencing factors as determined in environmental scanning within the framework of the extrapolative approach. Depending on the available resources, either the comprehensive approach of scenario technique or the prioritising technique of visioning can be used (Fig. 4).

The extrapolative approach shown on the left comprises several individual methods as follows.

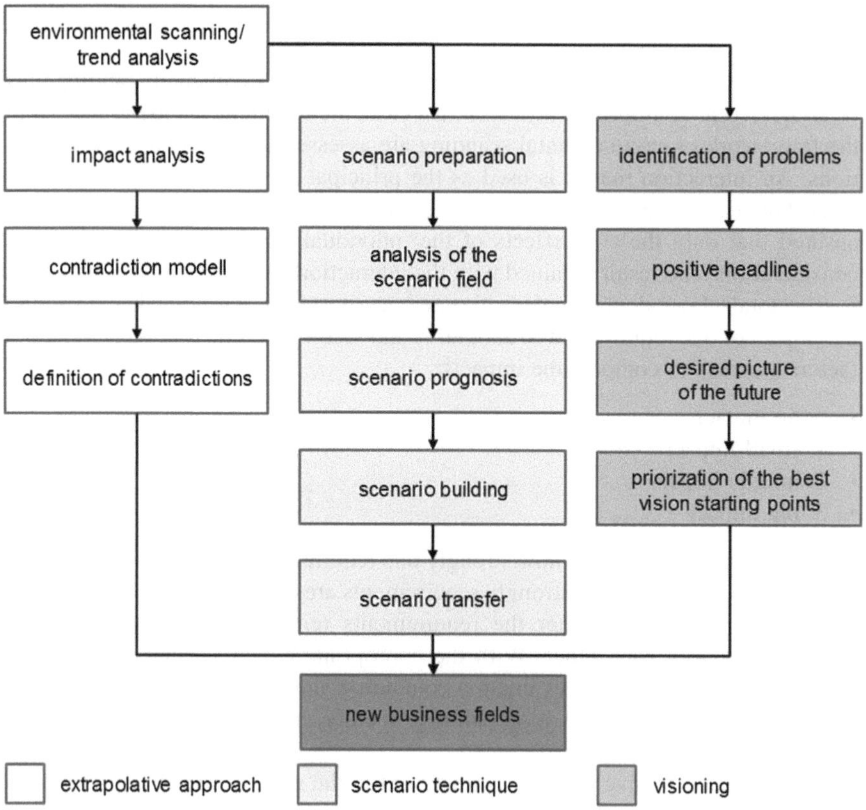

Fig. 4 Procedural structure for the derivation of search fields

The first step in the derivation of search fields is environmental scanning. With the aid of a STEEP analysis, trends for all areas of business are identified in the environment of a company, from which the effects and influencing factors are derived. As in the case of TRIZ, STEEP is an acronym and stands for "**S**ociety, **T**echnology, **E**conomy, **E**cology, **P**olitics". This means that trends are brought together in tabular form for all areas of business and for each of the five environment sectors. The environment sector Economy includes also the influences of customers, competitors and management as shown in Fig. 3. They are visualised in Fig. 3 to point out their importance. The information in this process was obtained from all available sources such as the Internet, reference literature and internal trend scouting data. The trends listed are then characterised by means of influencing factors, and the underlying requirements such as environmental protection, mobility and increasing energy demand are then determined from the perspective of the stakeholders. If shifts in the relevance of individual areas of business are found, these can additionally be taken into consideration. In order to reduce the amount of work required, it is advisable under certain circumstances not to perform the evaluation individually on all areas of business but to carry out clustering or prioritisation beforehand.

In the second step, the requirements determined in environmental scanning must be addressed in an impact analysis. In this case, the requirements identified within the framework of environmental scanning are assessed in relation to their interactions. An interaction matrix is used as the principal tool here. Since this work is carried out using requirements instead of quantifiable influencing factors, it must be ensured that only the core effects of the individual requirements are taken into consideration. The result obtained with the interaction matrix comprises active and passive totals. For calculating the active and passive totals, it is graded how strongly the impact of one requirement to another is and vice versa. The following valuation factors are used to compare the impacts:

- 0: no impact
- 1: small impact
- 2: middle impact
- 3: big impact

The active total indicates how strongly one requirement affects others, while the passive total describes how strongly requirements are influenced by others. It is not relevant in this case whether the requirements reinforce each other or are in contradiction with each other. With the appropriate expertise, individual assessments will give a statistical evaluation containing values that can hardly be differentiated from each other. A more suitable method here is joint assessment by a group of experts.

The active or passive totals are transferred in the next step to an active/passive diagram. The active/passive diagram is a representation of the portfolio that is flagged up by the relevant active and passive totals of the individual requirements. For each requirement, there is one point in the diagram, and the points that show the largest active and passive totals are identified as so-called key requirements.

The key requirements are displayed in the next step in an interaction model as shown in Fig. 5. The objective of this model is the representation of the interactions between the identified key requirements. The dashed-line arrows indicate mutual dependence, while a solid-line arrow symbolises requirements that contradict each other.

The central idea of the procedural structure prepared is to transfer the TRIZ core philosophy for the resolution of contradictions into the foresight process. Contradictions between the derived key requirements are identified, and the ways of overcoming these contradictions represent potential search fields. From the diagrammatic representation of the contradictions and the mutual dependences, triangular relationships are extracted, comprising two contradictions and one mutual dependence. The triangular relationship of the requirements "energy demand", "environmental protection" and "mobility" is used as an example.

The triangle is transferred into the WOIS bubble model (see Fig. 5). One of the two key requirements in the mutual dependence, in this case "energy demand", is used as a lead value. This means that, if there is an increase in this value, the

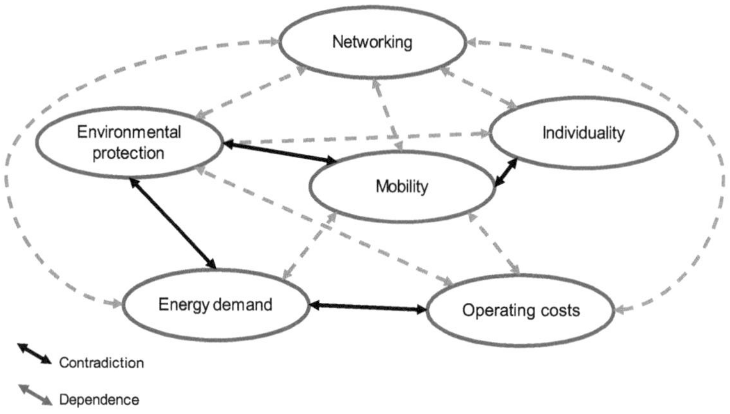

Fig. 5 Interactions of the key requirements

dependent target factor "mobility" will also increase, in contrast to the second target factor "environmental protection", which will decrease. This second target factor is accordingly in contradiction with the first target factor and the system parameter. These interactions are illustrated by the arrows between the bubbles.

In order to resolve the contradictions shown in the WOIS bubble model, the 40 inventive principles are applied in the classic TRIZ doctrine. This idea is now also transferred to the foresight field by identifying contradictions between the key requirements. Overcoming these contradictions represents a potential search field. Adapted "foresight principles" can be used to this end. These correspond in some cases to the classic innovation principles, and in other cases they are newly developed principles.

With the aid of the principles, impulses are identified on the basis of the data in environmental scanning that are used in the resolution of the contradictions. Impulses are described as brief descriptions of trends or individual aspects of a trend. The impulses found are clustered as search fields (Fig. 6).

The retropolative approach completes the derivation of search fields beside the extrapolative approach. It consists of scenario technique and/or visioning.

The scenario technique is used in a classical manner with the individual step scenario preparation, analysis of the scenario field, scenario prognosis, scenario building and scenario transfer as it is taught by J. Gausemeier (Gausemeier et al. 2009).

With the aid of visioning, desired visions of the future are created. These are drafted starting from the critique of the present or fears for the future formulated as headlines. The best vision nuclei derived from the desired visions of the future are the filtered out.

For the identified search fields, diagrammatic preparation of the data is possible in the form of a radar. This shows a time-based classification, in other words whether further research must fundamentally be carried out on a search field or whether the commercialisation phase has already been reached. This classification

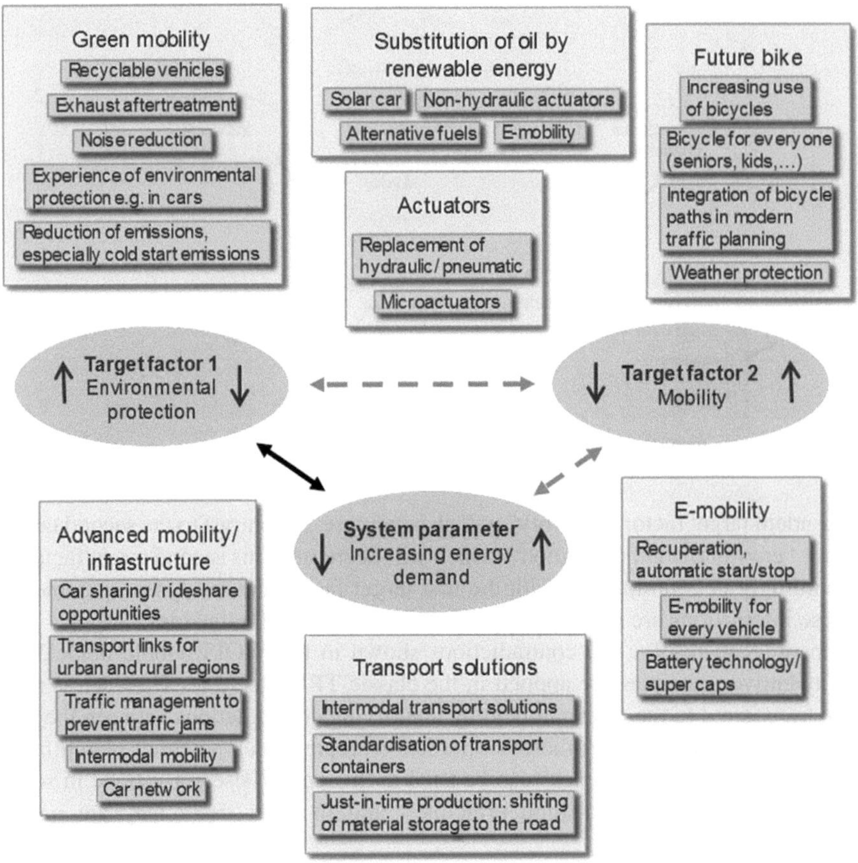

Fig. 6 Search fields in the contradiction between environmental protection and mobility

is comparable with the phases of the innovation process as presented originally. Further research is carried out in the fuzzy front end and the potentials of a technology, for example, cannot yet be foreseen. In addition to the time-based classification, the significance for the immediate company can be shown in a second dimension.

4 Procedural Structure for Development Directions

In addition to the procedural structure for deriving search fields, a further procedural structure was designed for development directions. Development directions are generated within a search field and contain the actual solution space. Development directions can be developed either on the basis of a top-down approach, in other words drawing from the future backwards, or through a bottom-up approach.

For the top-down approach, in other words where development directions are derived starting from the search field, the procedural structure can essentially be adapted in order to derive search fields. In this case, environmental scanning is first carried out for a particular search field. The requirements thus identified are assessed in terms of their interaction and transferred into the active/passive diagram. Key requirements can thus be derived for which contradictions can be defined in a similar way to the procedure for search fields. The contradictions are used as the starting point for deriving future functions and requirements.

The bottom-up approach draws on existing products or systems in the solution space. In this case, development directions are displayed using various TRIZ methods. Various methods and tools with different objectives are available to this end, which are taken from the classic TRIZ doctrine or further developments such as WOIS, I-TRIZ or Gen3. For each objective, a method was selected that was most suitable for the problem described as "identify development directions". One part of the methods taken from preliminary selection is later compiled in a procedural structure. The following objectives can be implemented with the aid of TRIZ methods:

- History analysis: With the history analysis of directed evolution from I-TRIZ, the individual development steps for a technical system can be displayed without a specified time period. It additionally incorporates the upper and lower system, technical boundaries in the past, the history of resources, the development of organisations included, the history of user markets, problems that have arisen and concepts that have been rejected (Zusman and Zlotin 2001).
- Product analysis/S-curve analysis: Each development proceeds along an S curve. S-curve analysis is used to determine the current position of a technical system on the S curve. It can thus be established how the system will develop in the future. By means of product analysis in accordance with WOIS, the system is separated into its individual components, and the position is determined for each individual component (Linde et al. 2008).
- Evolution laws: Developments can be shown with the aid of laws by using the development laws in accordance with WOIS. A distinction is made between stages, factors, steps, laws, strategies and phases. WOIS offers the advantage that an analysis is completed at the component level. It is also possible, on a similar basis to directed evolution, to show system weaknesses and hazards arising in the future as well as to identify methods and conditions that are required for the realisation of individual steps (Linde et al. 2008).
- Determination of overall objective: Overall objectives can be determined using the method for determination of overall objectives in accordance with WOIS. Moving away from the component level, individual overall objectives are defined for functions, actions and customer groups. Interactions between the components are also identified (Linde et al. 2008).
- System and system environment: The system environment is analysed with the aid of WOIS. In particular, this makes it possible to clarify the interactions between individual components or with the environment, the transfer of similar

functions to different components and extreme users and environmental conditions (Linde et al. 2008).

- Process analysis: Processes are characterised with the aid of process analysis in accordance with WOIS. The sequence and temporary dependences of the individual process steps in development and production are considered. Interpretation is focussed on particularly critical subfunctions (Linde et al. 2008).
- Development front analysis: The development front can be analysed with the aid of development front analysis in accordance with WOIS. The system is subdivided into its individual functions in accordance with a morphological procedure. For each subfunction, searching is then carried out for implementation structures. In addition, a targeted search is carried out for analogous structures specifically outside the core market sectors (Linde et al. 2008).
- Trend screening: For the analysis of trends, the trend screening function of WOIS can be used. The user is presented with a list of societal and technological trends. The user transfers these to the technical system and develops ideas for fulfilling objectives in terms of benefits or outlay by changing the lead values. The focus of the user is specifically on the identification of requirements and functions (Linde et al. 2008).
- Main parameters of value: For analysis of the added value characteristics of a technical system, the method main parameters of value from Gen3 can be applied. On a similar basis to development front analysis, functions and characteristics are defined that lead the customer to make a purchase. An analysis is then carried out for each function to determine which technical value is responsible for the fulfilment of the functions (Gen3 Partners 2014).

The methods described are essentially suitable for identifying development directions. However, a selection must still be made from the entire range of methods as a function of the customer or product portfolio, since many methods can only deploy their advantages for certain applications. The method that is suitable for a certain set of tasks can be found in Table 1.

Schaeffler is a business-to-business company and produces not only individual components but also complex technical systems. The appropriate procedure for the bottom-up approach is therefore one with history analysis, product analysis, the application of evolution laws and analysis of the development front, together with incorporation of current trends. For individual components or technical systems with very few components, product analysis should ideally be replaced by an S-curve analysis in accordance with the classic TRIZ doctrine.

In conjunction with the top-down approach, the procedural structure is as shown in Fig. 7. If Schaeffler wishes to make inroads into a new search field, future development directions are defined using the top-down procedure. If the aim is to develop existing products further and identify new development directions, on the other hand, the bottom-up approach is used.

Table 1 Guide for selection of methods for the derivation of development directions

	Business to business		Business to consumer	
Method	Technical system	Individual component	Technical system	Individual component
History analysis	Suitable	Suitable	Suitable	Suitable
Product analysis	Suitable	Not suitable	Suitable	Not suitable
Development laws	Suitable	Suitable	Suitable	Suitable
Determination of overall objective	Only suitable under certain conditions	Only suitable under certain conditions	Suitable	Suitable
System environment	Only suitable under certain conditions	Only suitable under certain conditions	Suitable	Only suitable under certain conditions
Process analysis	Suitable	Only suitable under certain conditions	Suitable	Only suitable under certain conditions
Development front analysis	Suitable	Suitable	Suitable	Suitable
Trend screening	Suitable	Preferably only for technological trends	Suitable	Preferably only for technological trends
Main parameters of value	Only suitable under certain conditions	Only suitable under certain conditions	Suitable	Suitable

5 Conclusion

The classic foresight process mainly encompasses intensive trend screening and its interpretation. This does not allow the systematic derivation of search fields or development directions. The procedural structure created starts with trend searching and its interpretation. From this starting point, the underlying requirements are analysed. The idea of generating contradictions between the identified requirements is based on the classic TRIZ doctrine and offers a high level of novelty in the search fields derived as a result. The essential challenge lies in overcoming the contradictions. This difficult task is supported in a systematic manner through the use of the classical TRIZ adapted from foresight principles.

The procedural structures derived using the approaches drawing from the future backwards (top-down) and starting from an existing solution (bottom-up) are also applicable, using the selected methods and tools, for small- and medium-sized enterprises with a justifiable level of work.

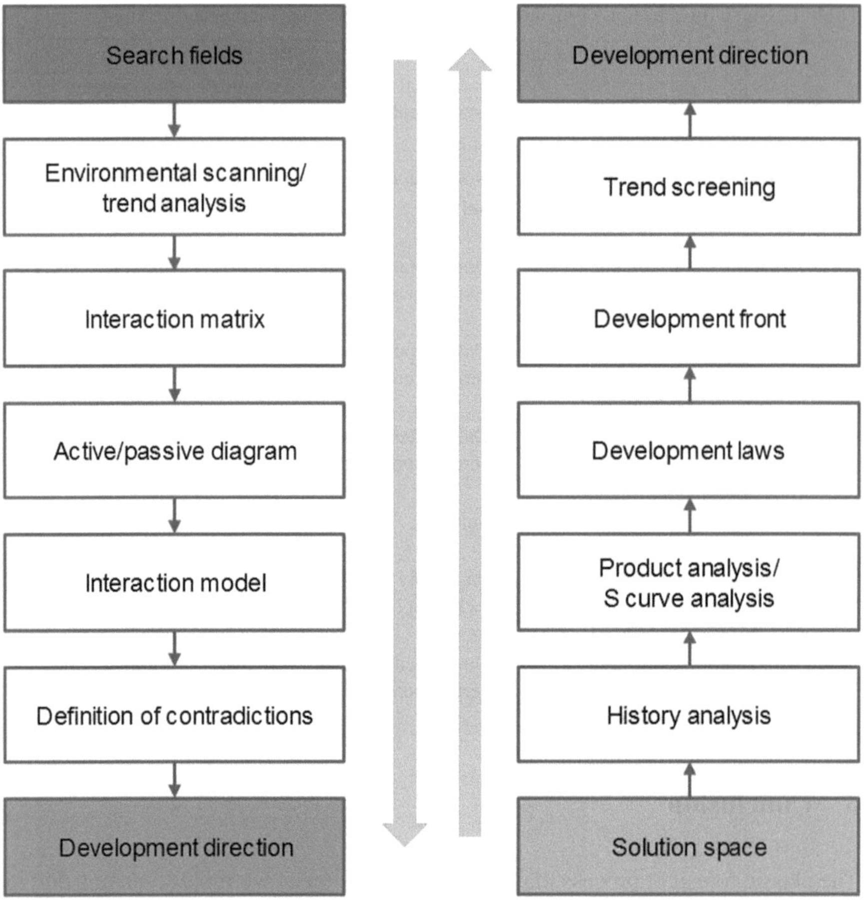

Fig. 7 Procedural structure for the derivation of development directions

References

Gausemeier, J., Plass, C., & Wenzelmann, C. (2009). *Zukunftsorientierte Unternehmensgestaltung—Strategien, Geschäftsprozesse und IT-Systeme für die Produktion von morgen*. München: Carl Hanser.

Gen3 Partners: What are main parameters of value (MPV's)? www.gen3partners.com. Accessed 28 Nov 2014

Koen, P., et al. (2001). Fuzzy-Front End, a.a.O., S. 2, Abb. 1.

Linde, H., Herr, G., & Rehklau, A. (2008). *WOIS- Widerspruchsorientierte Innovationsstrategie-Philosophie und Technologie zur Innovation der Innovationsprozesse*. Coburg: WOIS Institut.

Schaeffler Technologies AG & Co. KG. http://www.schaeffler.de/content.schaeffler.de/de/innovation/culture_innovation/innovation_process/innovation_process.jsp. Accessed 28 May 2015.

WOIS Institut: Strategische Orientierungsmittel, 2008.

Zusman, B., & Zlotin, A. (2001). Directed evolution—Philosophy, theory and practice, Farmington Hills. Available via Ideation International Inc. http://www.trizmantra.com/Learning%20Resource%20files/DirectedEvolutionBook.pdf

Five-Step Method for Breakthrough

Vladimir Petrov

Abstract The chapter presents a TRIZ-based algorithm to support idea generation in inventive problems. The method differs to all known algorithms including classical ARIZ, and it requires smaller amount of time to learn it and to apply while retaining the same amount and quality of ideas.

Forty years of experience of TRIZ application, teaching, and development resulted in this five-step easy-to-use roadmap.

The algorithm is supposed to be used by engineers, researchers, students, inventors, and people solving complex problems in business, government, and social environments, as well as everyday problem solvers.

1 Introduction

This chapter presents a new algorithm for the guided brainstorming in inventive problems of various backgrounds. The algorithm is based on the theory of inventive problem solving (TRIZ) developed by Genrich Altshuller (Altshuller 1984, 1987, 1999; Altshuller et al. 1999; Salamatov 1999). More specifically, the method is derived from its central part, which is called algorithm of inventive problem solving (ARIZ). ARIZ is a step-by-step program for the analysis and solution of inventive problems. The first version of the algorithm appeared in 1956 (ARIZ-56). Here is the list of all versions released by Altshuller and his followers: ARIZ-59, ARIZ-61, ARIZ-62, ARIZ-63, ARIZ-64, ARIZ-65, ARIZ-71, ARIZ-71B, ARIZ-71C, ARIZ-77, ARIZ-82, ARIZ-82A, ARIZ-82B, ARIZ-82C, ARIZ-82D, ARIZ-85A, ARIZ-85B, and ARIZ-85-C. The latest modification of ARIZ includes three basic components: **program, information support**, and **methods to control the psychological factors**.

As known, **ARIZ** consists of a sequence of operations for the following functions: exposure and solution of contradictions (see the basic sequence of ARIZ),

V. Petrov (✉)
Innovation Technology, Israel
e-mail: vladpetr@013net.net

L. Chechurin (ed.), *Research and Practice on the Theory of Inventive Problem Solving (TRIZ)*, DOI 10.1007/978-3-319-31782-3_8

analysis of the initial situation and selection of the problem to be solved, synthesis of the solution, analysis of the obtained solutions and selection of the best option, development of obtained solutions, documenting the best solutions, and generalization of these materials to improve the ideation method for approaching other problems. The structure of the algorithm and the mechanism for its implementation are based on the laws and regularities of engineering systems evolution.

The **information support** is provided by a **knowledge database**, which includes a system of *standards for the solution of inventive problems*, *engineering effects* (physical, chemical, biological, mathematical, and particularly geometric—the most developed effects at the present day), *techniques for the elimination of contradictions* (*inventive principles*), and *methods for the application of natural and technological resources*.

Methods for the control of psychological factors are necessary due to the fact that as ARIZ is not a computer program, the problems are being solved by humans. Therefore, a problem solver is influenced by psychological inertia, and it is necessary to control this phenomena. Furthermore, these methods assist creative imagination that is necessary in complex inventive problem analysis.

However, ARIZ turned out to be as powerful as demanding analytical tool. To study ARIZ, one needs at least 100 h and around 80 hands on exercises. Another limitation is the application field, and ARIZ was designed to solve technical problems only. Multiple reviews report a sad fact that the practice ignores ARIZ (Moehrle 2005; Ilevbare et al. 2013; Chechurin 2015).

Forty years of author's TRIZ practicing, teaching, discussing, and development resulted in formulation of a new version of inventive problem analysis roadmap that requires 4–8 h of training while being applicable for not only engineering domain. At the same time, the approach retains the valuable essence of TRIZ.

The five-step method was first introduced in 1977 (Petrov 1977) as the attempt to assist the growing difficulties of learning ARIZ. Indeed, ARIZ-1977 introduced bigger amount of particular steps, for example, the contradiction analysis was given the physical contradiction formulation and elimination technique. The first version of the method was called "ARIZ logic" since it navigates the user through it. There were some changes in the approach itself, and the biggest are listed below:

1. ARIZ steps are aggregated. Only administrative-engineering contradictions, ideal final result, physical contradictions, and solution.
2. The notation of contradictions was changed (from "engineering," "physical") to extend the ideation support to nontechnical fields.
3. The links between all types of contradictions are shown. Graphical model is suggested.
4. Physical contradiction analysis means many levels of depth. The contradiction is in-depth analysis until the key cause is revealed.

The new algorithm has been discussed and improved since 1977 in the enormous amount of seminars and consulting meetings by students, clients, and colleagues. The first English version appeared in 2004 (Petrov 2004).

Since its earlier days, the method has been used in various fields of engineering, business, management, etc. For example, shipbuilding, aviation and space, printing, various automatic control devices, quality control devices, robotics, military industry, communication and radio engineering, biotechnology, medicine and medical devices, sports and sports equipment, information technology, methods for improving strategies and tactics of various management processes, etc.

The remainder of the chapter is organized as follows:

Part 2 provides the whole algorithm at a glance and defines contradictions and the way to use them.
Part 3 introduces the concept of ideal result, its definition, and properties. Examples illustrate the concept.
Part 4 presents the detail problem solving process chain.
Part 5 describes the interconnection of all types of contradictions and ideal result. The graphical scheme of these links is given.

2 Five Steps

2.1 The Sequence of Steps

The method comprises five steps (Fig. 1):

1. Problem description
2. Defining the parameters of the system
3. Determining the direction of solutions
4. Determining the properties of the system
5. Solution

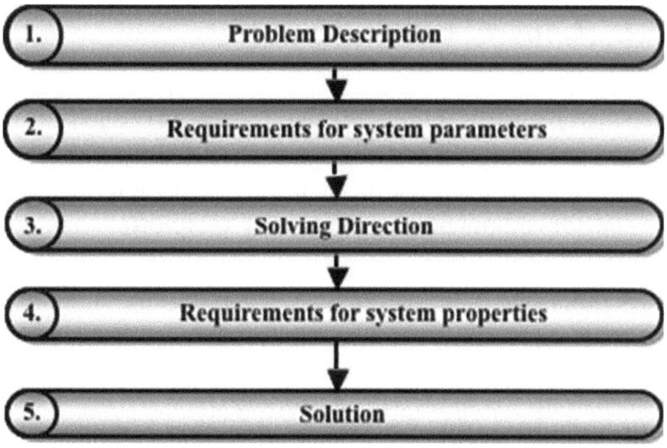

Fig. 1 The five-step method

2.2 *What Are the Steps and How to Use Them*

1. Problem description. Description is made in an arbitrary form, while it is obligatory to indicate:
   1.1. The purpose of the system (the main system function)
   1.2. The composition of the system (list all system components)
   1.3. The disadvantage of the system or what needs to be improved
2. Defining the system parameters.
   2.1. If there is a disadvantage, identify what good does the system do.
   2.2. If system improvement is necessary, indicate what bad result might accompany the system improvement.
3. Determining the direction of solutions.
   3.1. The solution direction is determined by an ideal result to be achieved.
4. Determining the system properties.
   4.1. Determine what property the system should have to eliminate disadvantage and what property the system should have that does not worsen the good properties of the system. Usually these are diametrically opposed requirements (properties).
5. Solution.
   5.1. Separate opposite properties, i.e., solve the problem.

   Specifying requirements to the system leads to the contradictions identification.

3 The Concept of Contradictions

3.1 *General Concepts*

A variety of different means have been developed to satisfy human needs. But the needs grow significantly faster than our ability to satisfy them, and this in its own way serves as a source of technological progress.

The development of a new entity more often than not entails the improvement of some set of technological parameters of a system.

Complicated inventive problems require a nontrivial approach because the improvement of one system parameter leads to the undesirable deterioration of another. The **contradictions** arise.

The solution of problems with the method constitutes a sequence for the exposure and solution of the contradictions and the reasons that produce the given contradictions, as well as their elimination by the use of the knowledge base. In this manner cause and effect relationships are determined—the essence of which is the intensification and aggravation of contradictions.

For this purpose, the methods consider three levels of contradictions. Altshuller named them respectively: *administrative*, *technical*, and *physical*.

3.2 The First Level Contradiction

The **first level contradiction (C-1)** is the contradiction between the expressed need and the ability to satisfy that need.

This contradiction lies on the surface. It is viewed by everyone. It is easy to define. These contradictions are often produced by administrators or customers; therefore, Altshuller called it *administrative contradiction*. It is expressed in the following possible manner:

- "It must be completed immediately, but it is unknown how."
- "Some kind of system parameter is faulty, so it must be fixed."
- "It is necessary to eliminate the shortcoming, but we don't know how."
- "There is spoilage in the production of wares, but the reason is unknown."

In this manner, the first level contradiction is expressed in the form of a **harmful effect (HE)**—something **negative** or the **necessity to create something new** by unknown means.

The first level contradiction is expressed as:

1. **Improvement (I)**—it is necessary to improve something, but how to do it is unclear.
   Is necessary to create parameter **A—good.**
   Then schematically denote that the **first level contradiction (C-1)** in case of **improvement (I)** is only a *good* parameter **A**.

$$\mathbf{C\text{-}1(I)} : \mathbf{A}$$

2. **Undesirable effect (UE)**—something is **bad**.
   Parameter **B—inadequate (bad).**
   Bad parameter conventionally denoted by **anti-B**.
   Then schematically denote that the **first level contradiction (C-1)** in case of **undesirable effect (UE)** is only a *bad* parameter **B**.
   Then schematically denote that the **first level contradiction (C-1)** in case of **harmful effect (HE)** is only a *bad* parameter **anti-B**.

$$\mathbf{C\text{-}1(UE)} : \mathbf{anti\text{-}B}$$

Problem 1. The Bus
The bus should transfer more passengers. What can be done?

This is the typical **first level contradictions**.

Type of the **first level contradictions**—improvement (I).

The **first level contradictions**: increase **capacity** (**A**).

Designing novel systems implies improvement of the system's technical parameters.

Complex inventive problems include contradictions, where improvement of parameters of the system worsens other parameters excessively.

3.3 The Second Level Contradiction

The **second level contradiction (C-2)**—this is a contradiction between specific parts, qualities, and parameters of a system. The second level contradiction arises during the improvement of one part (quality or parameter) of a system at the expense of the inadmissible deterioration of another. It reveals the reason for the appearance of the first level contradiction by intensifying it. More often than not, several second level contradictions (C-2) lie at the heart of the first level contradiction (C-1).

As a rule, by improving certain characteristics of an entity, we dramatically worsen others. Usually it is necessary to search for a compromise, that is, to sacrifice something.

The second level contradiction arises as a result of the disproportionate development of different system parts (parameters). When there are a significant number of changes to one of the system parts (parameters) and a sharp "lag" in development of another (other) of its parts, the situation arises in which quantitative changes from one side of the system (parameter "A") act in contradiction with others (parameter "B"). Solution of this kind of contradiction often requires the qualitative change of the technological system. This is manifested in the law of transition from quantitative to qualitative changes.

Altshuller considered contradictions in technical systems and called this kind of contradiction as "technical (or engineering) contradiction." Since modern TRIZ is applied to many kinds of systems including business and social systems, we prefer to use the more general term "second level contradiction."

For example, increasing the speed requires an increase in power; power increase leads to an increase in energy consumption, which in turn leads to an increase in size and weight.

Thus, the second level contradiction (C-2) is presented as follows:

- Speed—power
- Power—energy consumption
- Energy consumption—size and weight

Other examples.

- We want to make the object a stronger, but it gets heavier.

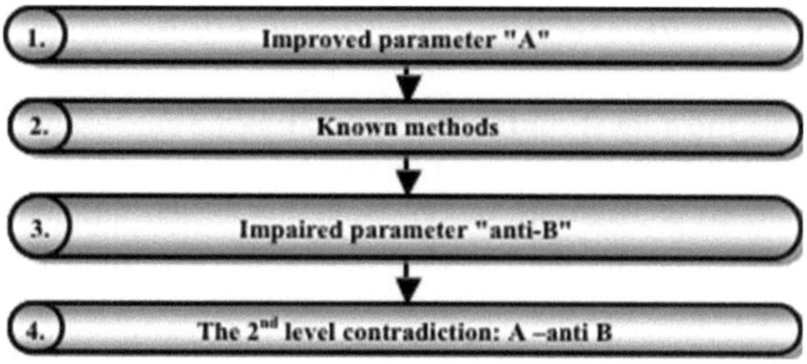

Fig. 2 Sequence for the second level contradiction identification

C-2: stronger—heavier.

- The company hires more employees to give better customer service, but its operating cost increases.
  C-2: better customer service—cost increases.
- We make information systems more secure, but the user interface is complicated and difficult to use.
  C-2: more secure—difficult to use.

3.3.1 The Sequence for the Second Level Contradiction Identification

The second level contradiction is defined as the result of answering the following questions (Fig. 2):

1. What we need to improve (parameter "A")?
2. Are there any known methods for such improvement?
3. What would be inadmissible in such an improvement (parameter "anti-B")?

   (a) Improvement in parameter "A" leads to a deterioration of the parameter "B."

   Continuation of the above example.

Problem 1. The Bus (Continuation)

1. What we need to improve (parameter "A")?
   - Increase the bus capacity (A).
2. Are there any known methods for such improvement?
   - Increased passengers' capacity requires a larger bus.
3. What would be inadmissible in such an improvement (parameter "anti-B")?

   (a) Improvement in parameter "A" leads to a deterioration of the parameter "B."

   C-2: large **capacity** (**A**)—low **maneuverability** (**anti-B**).

3.4 The Third Level Contradiction

The **third level contradiction (C-3)** is the coexistence of opposite properties (e.g., physical) in a certain part of a system.

This is necessary to determine the reasons producing the second level contradiction, i.e., the third level contradictions constitute the further intensification of the contradiction. Amplification (intensification) of contradictions can be continued to an even greater degree for the exposure of the contradiction's initial cause. The formulation of the third level contradictions sounds unaccustomed and even unorthodox—**some part of the system should be located simultaneously in two mutually exclusive states**:

- *Cold* and *hot*
- *In motion* and *motionless*
- *Long* and *short*
- *Flexible* and *rigid*
- *Electrically conducive* and *non-conducive*
- *Good* and *bad*
- *Thick* and *thin*
- *Complex* and *simple*
- *Want* and *do not want*
- *Love* and *not love*
- *Be* and *not to be*
- Etc.

The study of reasons that produce the second level contradictions in technological systems as a rule leads to the necessity of exposing contradictory physical properties of a system. Therefore, Altshuller gave them the name physical contradictions.

For clarity, the third level contradiction can be represented in the form of unusual interval.

The interval representing a zone "A" is presented in the form of inequality

$$\mathbf{a} < \mathbf{x} < \mathbf{b} \tag{1}$$

The graphical representation is given in Fig. 3.

The zone is used to search for the optimal solution, a compåromise.

TRIZ uses it in a completely different way.

First, define which property (**p**) should be the best for the system with parameter **A**. Denote this property as **P**. Next, define which property (**p**) should be the system to parameter **B** was the best. Denote this property as **anti-P**.

Such conflicting action is presented in the form of inequality

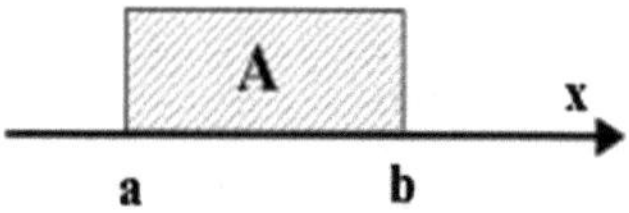

Fig. 3 The interval

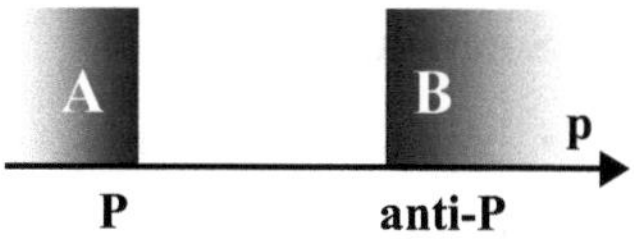

Fig. 4 Graphic representation of the third level contradiction

$$\mathbf{P > p > anti\text{-}P} \tag{2}$$

The physical contradiction can be presented as an unusual "interval." The graphical representation is given in Fig. 4.

We continue the examination of Problem 1.

Problem 1. The Bus (Continuation)

We will formulate the third level contradiction (C-3).

C-3: The bus should be **larger** in order to carry more passengers, and it should be **small** to maintain maneuverability.

In this manner, the three types of contradictions we have examined form a chain: the **first level contradiction (C-1)**—the **second level contradiction (C-2)**—the **third level contradiction (C-3)**, which determines the cause and effect relationships in the technological system under examination.

$$\mathbf{C\text{-}1 \rightarrow C\text{-}2 \rightarrow C\text{-}3} \tag{3}$$

Conflicting properties should be resolved, i.e., we obtain a **solution (S)**.

$$\mathbf{C\text{-}1 \rightarrow C\text{-}2 \rightarrow C\text{-}3 \rightarrow S} \tag{4}$$

In the simplest case, contradictory properties can be resolved:

- *In space*
- *In time*
- *In structure*
- *By condition*

Problem 1. The Bus (Continuation)

The contradiction can be separated:

- *In structure*, by making the bus dynamic or flexible.
  Solution:
  - Two buses are joined with a bendable joint (as an *accordion*).

- *In space.*
  Solution:
  - A bus is put on top of another bus—*double-decker bus.*
- *In time.*
  Solution:
  - A lot of small buses.
- *By condition.*
  Solution:
  - The bus is expandable. At the peak hours, the bus is expanded and can take a lot of passengers. Bus concertinas when few passengers. The bus is made telescopic or accordion-like.

3.5 Features of the Five-Step Method

In conventional design, when improving certain characteristics of an entity, we frequently dramatically worsen others. Then it is necessary to search for a compromise, that is, to sacrifice something. TRIZ and the five-step method are different—you use your understanding of the three levels of contradiction to find a solution that gets rid of the problem, not a compromise that is only a partial solution.

Sometimes just the identification of the level 1, 2, or 3 contradiction can lead directly to the solution. In other cases, the level 2 or 3 contradiction definition can lead to a refinement of the level 1 contradiction statement, etc. Do not be surprised if several iterations of thinking through the sequence of level 1, level 2, and level 3 are necessary to get a useful definition of the problem.

4 The Path to the Ideal

The solution of mathematical problems and problems "by quick-wittedness" is often achieved by the method "reversing the process." The essence of this method lies in the fact that in order to solve a problem, you need to start with the end result. The answer is to determine the final result. Clarifying this, the road to the beginning, i.e., the solution of the problem, is "paved."

It would be misleading to undertake the solution of technological problems in the same way. But how can one find out the answer?

Indeed, during the solution of technological problems, the end result is unknown; however, it is still possible to go on. It is possible to imagine the ideal of a developed construction—this ideal construction is the **ideal result (IR)**.

Altshuller called it **ideal final result (IFR)**.

IR is the beacon to which one should look during the solution of problems. The closeness of the received solution to the ideal determines the level and quality of the solution. IR is the solution we would like to see in our dreams, carried out by fantastic creatures or means (magic wand), for example, a road that exists only where it touches the wheels of transport vehicles.

Main Properties of IR

The ideal result contains:

Conflicting parameters comprising the contradiction are both good.

System is not more complicated.

All harmful effects are removed.

Every operation and result appear at the correct time and place.

No outside intervention is needed in order to perform all operations.

One of the basic traits of an IR is that it should **appear only at the moment when it is necessary to complete the necessary action, and, furthermore, at that time it should undertake 100 % of the rated working load.**

This trait has been long known to us in fairy tales—"magic tablecloth," etc.

Many examples can also be found in real life: retractable and folding objects, folding and attachable furniture (tables, couches, beds, etc.), and inflatable objects (boats, emergency vests, mattresses, pontoons, etc.).

Example 1 For the rescue of people during emergency aircraft landings in water, English engineers developed a safety device that constitutes pontoons which automatically inflate with compressed air.

IR of a means of transport is when it does not exist, but nonetheless the load is transported (the load "itself" travels in the necessary direction with the necessary speed).

Example 2 Automobile seat belts must be changed periodically. The concern has been expressed that the material may weaken. Belts have been invented that show by their appearance when they need changing.

Example 3 A layer of colored paint is applied to the tread of a protector and the number of kilometers traveled by the automobile before attrition of the applied layer is recorded. This method for evaluating the wearability of the tire is simple and useful during testing of the longevity of new types and constructions. This method can be used during the inspection of tires for their replacement.

Example 4 Window glass needs cleaning. To undertake this operation on high, large shop windows is somewhat labor intensive. If the shops are "glazed" with Lavsan film, then during a gust of light wind the film **itself** clears dust from the window. This film is transparent, light, and unharmed by hydrofluoric acid fumes. For the "glazing" of windows with this type of film, lightened frames can be used.

5 The Path to the Idea of Solution Concept

Having examined the basic concepts of the method—IR, first (C-1), second (C-2), and third (C-3) level contradictions—we can easily imagine the stages in the exact formula of a technical problem.

The definitive basic procedure for the solution of problems with the method can be represented in the following manner:

$$\textbf{C-1}\rightarrow\textbf{C-2}\rightarrow\textbf{IF}\rightarrow\textbf{C-3}\rightarrow\textbf{S} \tag{5}$$

From the point of view of the method, the problem is formulated exactly when C-1, C-2, IR, and C-3 are revealed in accordance with the chain represented above (5). To formulate all its links, first of all reveal that which does not suit the "poser" of the problem in the given situation (first level contradiction—C-1) and what is faulty in the system (undesirable effect). What kind of requirements is necessary to expect from the system?

Thus, the second level contradiction (C-2) is determined. Then the system is expressed in such a way that the undesired effect is absent from it, yet the positive qualities are preserved. The result of expressing the system in this manner constitutes the formulation of the ideal result—IR. After comparison of the actual situation and the IR, obstacles to achieving the ideal result are revealed, reasons for the appearance of the obstacles are sought, and the contradictory qualities that appear in certain parts of the system (operative zones) and do not satisfy the requirements of the IR are determined. In this manner, the third level contradiction (C-3) is formulated, which constitutes the exact formulation of the problem.

6 Logic of the Five-Step Method

The logic of solving problems with the five-step method shows the interconnections between elements in the basic sequence (5):

$$\textbf{C-1}\rightarrow\textbf{C-2}\rightarrow\textbf{IF}\rightarrow\textbf{C-3}\rightarrow\textbf{S} \tag{6}$$

The **first level contradiction (C-1)** is described either as a need for the appearance of a *new quality* or *action* **A** (**positive effect**) or in the form of a **harmful effect** (**anti-B**), which is necessary to eliminate.

The **first level contradiction (C-1)** more often than not is expressed in the form of a **harmful effect** (**HE**), i.e., a parameter or requirement B in an undesirable, harmful, or insufficient condition that we label **anti-B**. It is represented in a diagram like this:

C-1 (HE) : anti-B

For the determination of the **second level contradictions (C-2)**, we expose two contradictory requirements of the system. We represent these requirements with the letters **A** and B. The intensified contradiction can be presented as a need for the improvement of characteristics satisfying requirement **A**, which leads to the unacceptable deterioration of characteristics satisfying requirement B (expressed as requirement **anti-B**). The undesired effect consists of requirement B. Or the contradiction can be formulated in reverse—the improvement of B at the cost of the deterioration of **A** (expressed as **anti-A**).

C-2: A – anti-B or **anti-A – B**

The formula of the **ideal result (IR)** should aim to eliminate the harmful effect (**anti-B**) while preserving the positive requirement **A**, that is:

IR: A, B

The **third level contradiction (C-3)** is determined by exposing contradictory properties **P** and **anti-P** (e.g., physical), which should be possessed by the element of the system that does not correspond with the requirements of IFR. For this, it is necessary to determine what property **P** the element should possess in order to preserve requirement **B**, i.e., in order to eliminate the undesired effect. Simultaneously, the same element should possess the contradictory property (**anti-P**) in order to preserve the positive effect **A**. In this manner, the element should possess the property **P** in order to satisfy the requirement **B** and quality **anti-P**, in order to preserve the requirement **A**.

C-3: P→A, anti-P→B

Further intensification of the contradiction takes place by means of exposing more deep (latent) properties $\mathbf{P_1}$, which are necessary for the creation (preservation) of the earlier exposed property **P**.

$$\mathbf{P \rightarrow P_1}$$

In some cases during the solution of complicated inventive problems, it is necessary to expose even deeper cause and effect relationships in the system. For this it is necessary to expose even more deeper properties $\mathbf{P_2}$, $\mathbf{P_2}$, ... $\mathbf{P}_n$. The next quality in the progression determines the reason for the emergence of the previous quality, i.e., that which is necessary for the fulfillment of the quality.

$$\mathbf{P_2 \rightarrow P_1}$$
$$\mathbf{P_3 \rightarrow P_2}$$
$$\ldots \rightarrow \ldots$$
$$\mathbf{P_n \rightarrow P_{n-1}}$$

In these cases several aggravated contradictions are exposed (**C-3$_1$, C-3$_2$, C-3$_3$, ... C-3$_n$**). This can be represented in a diagram:

$$\text{C-3}_1 : \mathbf{P \rightarrow P_1;\ \ anti\text{-}P \rightarrow anti\text{-}P_1}$$
$$\text{C-3}_2 : \mathbf{P_2 \rightarrow P_1;\ \ anti\text{-}P_2 \rightarrow anti\text{-}P_1}$$
$$\ldots$$
$$\text{C-3}_n : \mathbf{P_n \rightarrow P_{n-1};\ \ anti\text{-}P_n \rightarrow anti\text{-}P_{n-1}}$$

Solution of the problem (S) consists of the solution of the aggravated contradiction, for example, by means of separation of the contradictory properties **P** ... **P$_n$**.

$$\mathbf{S:}$$
$$\mathbf{P \mid anti\text{-}P}$$
$$\mathbf{P_1 \mid anti\text{-}P_1}$$
$$\ldots$$
$$\mathbf{P_n \mid anti\text{-}P_n}$$

Typical methods of separating contradictory qualities are indicated in the chapter.

The logical diagram of the five-step method for the solution of problems is shown in Fig. 5.

Logical diagram of the five-step method with in-depth analysis is shown fully in Fig. 6.

Problem 2. Gas Vessels[1]

It is necessary to transfer all of the gas from a transport vessel to two empty (working) vessels of cylindrical form. The capacity of each of the working cylinders is equal to one half of the capacity of the transport cylinder.

Two methods for transferring gas are known (Fig. 7).

In the first method (Fig. 7b), the transport cylinder connects directly with the working cylinder. In this case, an equal pressure is fixed in all the cylinders, and one half of the gas will remain in the transport cylinder. The second method (Fig. 7c) is much more complicated: Gas is pumped from the large cylinder into two others by means of a compressor. In this manner, it is possible to transfer all of the gas, but it is necessary to use special equipment—a high-pressure compressor.

---

[1] The problem was published first in Altshuller (1961), pp. 45–46: The chapter presents the original analysis of the problem.

Fig. 5 Logical diagram of the five-step method

C-1 (HE): **anti-B**
↓
C-2: **A – anti-B** or **anti-A – B**
↓
IR: **A, B**
↓
C-3: **P → A, anti-P → B**
↓
S: **P | anti-P**

- ***in space;***
- ***in time;***
- ***in structure;***
- ***by condition.***

Fig. 6 Logical diagram of the five-step method with in-depth analysis

C-1 (HE): **anti-B**
↓
C-2: **A – anti-B** or **anti-A – B**
↓
IR: **A, B**
↓
C-3: **P → A, anti-P → B**
↓
C-3$_1$: **P→P$_1$; anti-P→anti-P$_1$**
↓
C-3$_2$: **P$_2$→P$_1$; anti-P$_2$→anti-P$_1$**
↓
… **…**
↓
C-3$_n$: **P$_n$→P$_{n-1}$; anti-P$_n$→anti-P$_{n-1}$**

S:

P | anti-P
P$_1$ | anti-P$_1$
…
P$_n$ | anti-P$_n$

- ***in space;***
- ***in time;***
- ***in structure;***
- ***by condition.***

The problem consists of finding a method to fully transfer the gas from the transport cylinder to the working cylinder without the use of additional equipment (compressor).

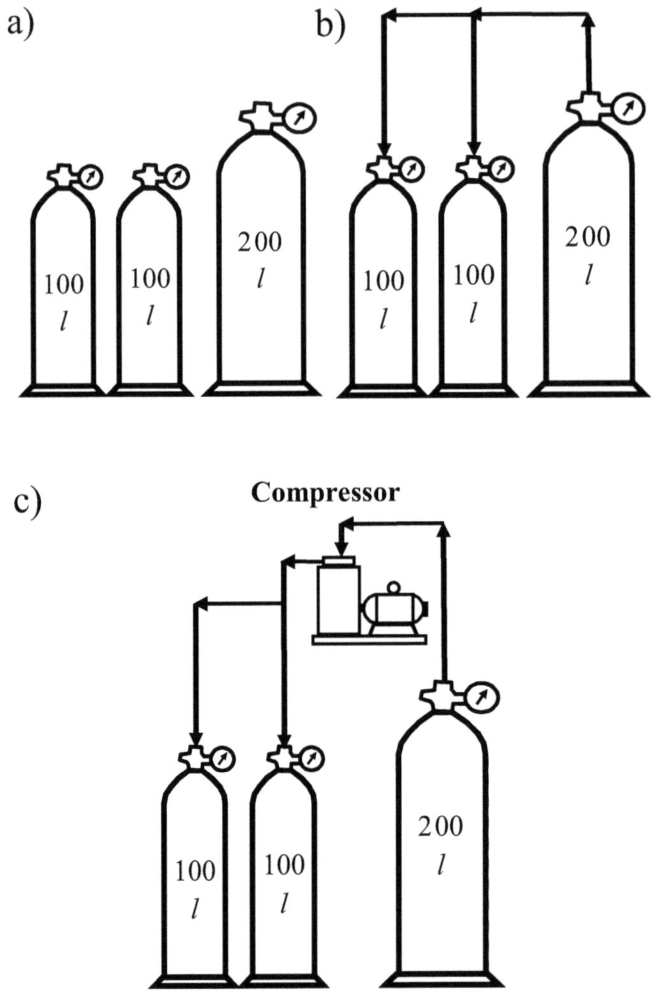

Fig. 7 Problem 1. Gas vessels' (a) initial situation, (**b**) direct connection of the transport cylinder with the working cylinder, and (**c**) connection by means of a compressor

This kind of problem is encountered during the "loading" of cylinders in deep-diving apparatuses at seaports. Compressed air is used for purging the cistern during surfacing of the apparatus.

6.1 Solution to the Problem

Short formulation of the problem: *Find a simple method of transferring all of the gas from one cylinder to two others.*

C-1:anti-B

The **first level contradictions (C-1)**: Part of the gas remains in the cylinder.
Type of *administrative contradiction*—**undesirable effect (UE)**.
Harmful effect: incomplete (**anti-B**) transfer of gas.
Determination of the **second level contradictions (C-2)**.
In the given example, the transfer of gas is possible *with* and *without* the use of a *compressor*:

1. With compressor

C-2_1: B–anti-A

C-2_1: The gas transfers completely (**B**), but as a result the system is complicated (**anti-A**).
C-2_1: Complete transfer of gas—complication.
All of the gas can be transferred from the transport cylinder to the working cylinder using a compressor, which complicates the system.

2. Without compressor

C-2_2:A–anti-B

C-2_2: The system is not complicated (**A**), but the gas transfers incompletely (**anti-B**).
C-2_2: Simplicity—loss of gas.
A simple method (direct connection) is used, but as a result one half of the gas is lost.
Selection of TC. We select TC_2, because this formula is based on the use of a simple method.

Note: Within this step we selected a method of transferring gas only through the direct connection of one cylinder with another.

Formulation of IR

IR: A, B

IR: The gas "itself" fully—(B) (with the same pressure and in the same quantity) transfers from one cylinder into two others, without the use (**A**) of additional equipment (compressor).
IR: Simplicity—loss of gas.
Formulation of the **third level contradictions (C-3)**.

$$\text{C-3} : \text{P} \rightarrow \text{A}, \text{anti-P} \rightarrow \text{B}$$

In order not to complicate the system, it is necessary to directly connect the cylinder with gas to an empty (working) cylinder, but this increases the general capacity in which the gas is located (decreasing its pressure), therefore not allowing the gas to be fully transferred.

C-3: In this manner the "extra" capacity (**P**) should exist, so that the system remains simple (**A**), and should not exist (**anti-P**), so that the gas transfers completely (**B**).

Note: Remember that the basic characteristic of gas is to occupy the entire available volume. Therefore, during connection of the working cylinders, gas expands and occupies the entire capacity of the cylinders, while the pressure decreases.

Formulation of **third level contradiction 1 (C-3_1)**.

$$\text{C-}3_1 : \text{P} \rightarrow \text{P}_1, \text{anti-P} \rightarrow \text{anti-P}_1$$

C-3_1: So that there is not an excess capacity (**P**), the working cylinder should not be empty (it should be filled) (**P_1**), but to provide space for the transfer of gas (**anti-P**), the capacity of the working cylinder should be empty (**anti-P_1**).

The connected cylinders should be filled to prevent the gas from expanding, but should not be filled (should be empty) so that they can be filled with the required gas.

Note: In this step we determined the exact formulation of the problem.

Solution of the problem.

The separation of contradictory properties can take place in space, in time, and by changing the system's structure, specifically changing the *aggregate condition.*

Therefore, the contradictory quality is that the working cylinder should be **full** and **empty** (**filled** and **unfilled**).

In Space This contradiction *is unsolvable.*

In Time Separation of the indicated contradictory qualities *in time* requires that the *substance* filling the working cylinder *gradually free space for the gas* entering from the transport cylinder and fill the freed space in the transport cylinder.

In Structure It only remains to explain what kind of substance should fill the working cylinder. For this, structural changes of the substance are used, changing the aggregate condition of the gas.

The substance inside the working cylinder is located in a gaseous state, which does not satisfy our conditions. That means it should be made solid or liquid.

Should the cylinder be filled with a solid substance? A solid monolithic substance does not possess the necessary qualities. In this manner we could ruin the cylinders. Of course, it is possible to fill the cylinders with sand or ice. That condition might solve the problem in principle, but it is not sufficiently effective. It remains to use liquid.

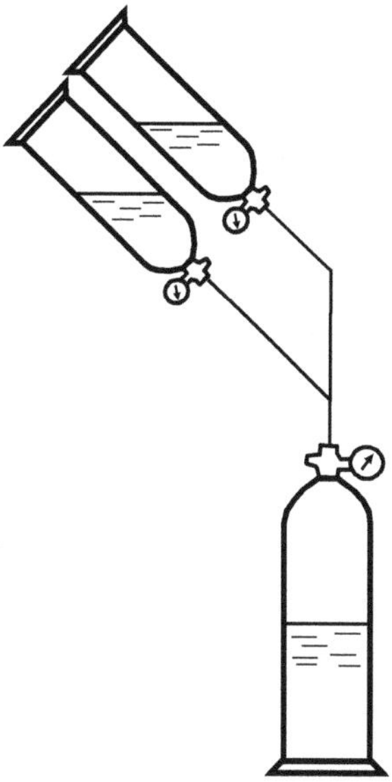

Fig. 8 Solution of Problem 2. Gas vessels

If the working cylinders are filled with a liquid excluding gas, placed above the transport cylinder, and connected by means of pipes, then the gas (fully and without a compressor) will transfer from the transport cylinder to the working one (Fig. 8).

Problem 4. Eye Needle

It is difficult to pass a thick thread through the small eye of a needle.

What is to be done?

C-1:A

The **first level contradictions (C-1)**: It is necessary to pass a thick thread through the small eye of a needle.

Kind of *the administrative contradiction*—**improvement (I)**.

Good parameter **A—ease of passage** (facilitate insertion of the thread).

C-2:A−anti-B

C-2: Ease of passage (A)—cloth spoiled (anti-B).

IR: A, B

IR: Ease of passage (A)—not cloth spoiled (B).

C-3:P→A, anti-P→B

C-3: A needle must have a **large eye** (**P**) to facilitate insertion of the thread [**ease of passage** (**A**)] and must have a **small eye** (**anti-P**) for convenient sewing [**not cloth spoiled** (**B**)].

$$\mathbf{S : P|anti\text{-}P}$$

S: By separating the contradiction **in time**.

The eye must be large while the thread is inserted and must be small during sewing. Inventive principle 15: dynamicity.

Solution 1 Robert K. Pace proposed that a needle be composed of two thin, spring steel wires of identical length (patent US 3,987,839).

The two wires are welded together at one end, twisted three quarters of a turn, and then welded together at the opposite end, after which the needle is polished and sharpened.

The resultant needle looks like an ordinary needle. When the needle is slightly unwound (by hand), a large slot is produced through which a thread is easily passed.

Upon its release, the needle returns to its initial shape and grips the thread.

S: By separating the contradiction **in space**, **structure**, **time**, and **dynamicity**.

Needle section—"eye" was separated from the needle.

Solution 2 Needle threader.

7 Conclusions

The packaging of classical TRIZ tools in the form of ARIZ algorithm meets serious difficulties in its practical applications. One is its complexity and the amount of time needed to learn and master it. Another is the limitation of the application or the algorithm by engineering domain mostly. The new algorithm presented in the chapter is based on TRIZ but seems to be easier to learn and apply while covering nontechnical application fields as well. The ideation process consists of a sequence of analytical steps for defining and eliminating contradictions or the causes of these contradictions, addressing the contradiction elimination with the help of knowledge database. In this manner, cause and effect relationship analysis framework is given, and the essence of which is the intensification and aggravation of contradictions. The presented five-step algorithm logic helps to reveal the key causes of a problem and to build cause–effect relationship model.

Acknowledgments The author is grateful for all the students, partners, colleagues, and clients for the feedback on the five-step algorithm. The author is especially grateful for Leonid Chechurin for his comments, discussion, and time spent on shaping the publication to the existing form.

References

Altshuller, G. S. (1961). *How to start inventing*. Tambov: Tambov Publishing (in Russian).

Altshuller, G. (1984, 1987). *Creativity as an exact science* (A. Williams, Trans.). New York, NY, London, Paris: Gordon & Breach Science Publisher.

Altshuller, G. (1999). *The innovation algorithm. TRIZ, systematic innovation and technical creativity* (p. 312). Worcester, MA: Technical Innovation Center, Inc.

Altshuller, G., Zlotin, B., Zusman, A., & Philatov, V. (1999). *Tools of classical TRIZ*. Southfield, MI: Ideation International Inc.

Chechurin, L. (2015, October 25–28). *TRIZ in science. Reviewing indexed publications*. TRIZfuture-2015 Conference Proceedings. Berlin.

Ilevbare, I., Probert, D., & Phaal, R. (2013). A review of TRIZ, and its benefits and challenges in practice. ResearchGate. http://www.researchgate.net/publication/257002814_A_review_of_TRIZ_and_its_benefits_and_challenges_in_practice

Moehrle, M. G. (2005). How combinations of TRIZ tools are used in companies—Results of a cluster analysis. *R&D Management, 35*, 285–296. doi:10.1111/j.1467-9310.2005.00390. http://onlinelibrary.wiley.com/doi/10.1111/j.1467-9310.2005.00390.x/abstract.

Petrov, V. M. (1977). The logic of ARIZ-77. Information for students.—L (manuscript, Russian).

Petrov, V. (2004, November 3–5). Logic of ARIZ. ETRIA TRIZ Futures Conference 2004. Florence (pp. 315–331). *The TRIZ Journal*. http://www.triz-journal.com/archives/2005/11/09.pdf, http://triz-summit.ru/en/204246/204247/

Salamatov, Y. (1999). *TRIZ: The right solution at the right time: A guide to innovative problem solving* (p. 256). The Netherlands: Insytec.

Part II
Case Study

TRIZ in Enhancing of Design Creativity: A Case Study from Singapore

Iouri Belski, Teng Tat Chong, Anne Belski, and Richard Kwok

Abstract This chapter investigated the ability of Theory of Inventive Problem Solving (TRIZ) to foster design creativity by infusing effective creative design strategies onto users. It explored the winning strategies of famous design innovators and engineering experts from the existing case studies and mapped these strategies onto the TRIZ-facilitated development of an anti-ram vehicle barrier carried out by a small engineering team from Singapore Technologies Kinetics. It has been discovered that the TRIZ tools of Situation Analysis, Method of the Ideal Result, 40 Innovative Principles and Principles of Separation engaged the project team in activities that resulted in at least three effective creative design strategies. During TRIZ facilitation the design team used (i) extensive situation analysis and task reframing, (ii) designing from fundamental principles and (iii) analogical thinking. These strategies allowed the team to replace originally poor design by a novel design that was successfully implemented and patented.

Keywords TRIZ • Creative design • Case study • Framing • Engineering design • Design techniques

Many authors have investigated the paths to creative ideas (e.g. Amabile 1983; Mednick 1962; Simon 1996; Howard et al. 2008; Akin 1990). Nonetheless, the role played by formal methods of idea generation in design creativity is still not entirely clear. As a result of a literature review, Shah et al. (2003) reached a rather negative conclusion on the effectiveness of ideation methodologies in engineering design:

> Despite several claims and much anecdotal evidence about the usefulness of these methods, little formal experimental evidence exists currently to indicate that these methods are effective in engineering design to generate concepts (Shah et al. 2003).

I. Belski (✉)
Royal Melbourne Institute of Technology, Melbourne, Australia
e-mail: iouri.belski@rmit.edu.au

T.T. Chong • R. Kwok
Singapore Technologies Kinetics, Singapore, Singapore

A. Belski
Swinburne University of Technology, Hawthorn, VIC, Australia

L. Chechurin (ed.), *Research and Practice on the Theory of Inventive Problem Solving (TRIZ)*, DOI 10.1007/978-3-319-31782-3_9

Providing experimental evidence in support of the influence of creativity methodologies on engineering design is somewhat challenging. Real engineering design tasks are complex and open-ended and are usually resolved under conditions that are different to laboratory experiments. Consequently, the outcomes of planned laboratory experiments may not always be generalised to real design conditions. It is posited that real design case studies that involve application of ideation techniques might be more appropriate than planned experiments for establishing the value of these techniques in design creativity.

This chapter presents a case study on the application of Theory of Inventive Problem Solving (TRIZ) by a team of engineers from Singapore Technologies Kinetics (ST Kinetics) for the development of a mobile anti-ram vehicle barrier (crash barrier). It attempts to practically establish the influence of TRIZ on the creativity of the team's solution by comparing designs proposed with and without TRIZ. Over 4 months of development, the project team suggested two different designs. Initially, the team proposed the crash barrier design without using any tools for idea generation. After the team members concluded that the original design was poor and unlikely to win a government tender, did they ask a TRIZ expert to facilitate a TRIZ session with the team. Over 2 days, the team applied the TRIZ tools of Situation Analysis (SA), Method of the Ideal Result (MIR), the 40 Innovative Principles and the Principles of Separation. The new design of a mobile crash barrier that was generated during these TRIZ sessions won the tender, was successfully implemented and deployed by the Singapore government, was patented and was also praised by Lee Kuan Yew, former Prime Minister and the founding father of modern Singapore (Lee 2005).

In order to evaluate the creativity impact of TRIZ in this case study, it has been hypothesised that an effective technique for idea generation helps a practitioner to 'think like a creative designer' and can infuse effective creative design strategies onto a user. The second design of the crash barrier was novel and was patented. Thus, it was expected that at least some of the TRIZ tools that were employed by the team during the TRIZ sessions were able to infuse effective creativity strategies onto the team members. This paper investigates the influence of TRIZ on design creativity and is specifically devoted to establishing the creative strategies that can be infused by the TRIZ tools of SA, MIR and 40 Innovative Principles.

1 Methodology

In order to establish the effective strategies that can enhance human creativity, the authors analysed the existing case studies that described practices adopted by the most creative engineering designers. Then, to find out whether the TRIZ tools led the team to the same creative strategies, a record of the team's TRIZ idea generation sessions was examined.

The information relevant to the team's design process has been acquired from two sources: design and patent documentation produced by the team and the diary entries made by the TRIZ facilitator that were made during and directly after TRIZ facilitation sessions. The TRIZ facilitator was not a part of the original design team. He attained sound knowledge in TRIZ (MATRIZ Level 3) and led the application of TRIZ to company projects for over 5 years. Being also a PhD candidate at the time, the TRIZ facilitator carried his research on corporate innovation as Action Research (Lewin 1947) and kept a comprehensive diary of his activities as they related to company innovation. Therefore, the TRIZ sessions with the crash barrier team were properly documented by him.

2 Creative Design Strategies of Famous Design Innovators

In order to obtain insights into design creativity, many researchers have explored the winning approaches of famous design innovators and engineering experts (e.g. Candy and Edmonds 1996; Cross and Cross 1996; Roy 1993; Maccoby 1991; Lawson 1994; Cross 2001a, 2003; Cross and Cross 1998). Analysis of these case studies and examination of the records of the TRIZ facilitation sessions assisted the authors in shortlisting the following three general strategies that the TRIZ tools were likely to infuse onto the crash barrier designers: (i) extensive situation analysis and task reframing, (ii) designing from fundamental principles and (iii) analogical thinking.

Many scholars have referred to the importance of these three strategies in creativity. Mednick (1962), for example, singled out the associative ability (analogical thinking) of individuals as the key to creative process. Simon (1996) postulated the importance of recognition, which is closely related to framing, in achieving innovative design solutions. A recent review of design and ideation processes (Howard et al. 2008) concluded that proper task analysis is essential in achieving creative solutions.

It is important to note that although the above-mentioned three general strategies were used by many famous designers, the way a specific strategy was realised by individual innovators differed significantly. For example, it has been reported that the strategy of task analysis and reframing (i) was realised by conducting the following activities:(a) Altering the initial constraint set (Akin 1990); (b) Reassessing the problem from the first principles (Cross and Cross 1996; Cross 2001b, 2003); (c) Re-evaluating the product's application (Cross 2001b); (d) Identifying the key question to be addressed (Candy and Edmonds 1996); (e) Immersing in the problem situation, studying it in depth and reflecting (Maccoby 1991; Roy 1993); (f) Changing the focus of task analysis from the systems level to the level of individual elements and then back (Maccoby 1991). Designing from the first principle (strategy ii) was usually fulfilled by deploying the fundamental principles of physics (Cross and Cross 1996; Cross 2003) and acquiring in-depth expertise (Roy 1993). Creative designs by analogical thinking (strategy iii) were achieved either by means of a mental transfer of technology from one application to another or by identifying analogies between the problem and various other products with similar functionality (Roy 1993).

The fact that the same creativity strategy can be realised by individuals through many different activities is of significant importance for the effectiveness of TRIZ and idea generation methodologies in general. This suggests that in order to be effective, an ideation methodology must engage a practitioner in activities that are capable of infusing sound creative strategies onto the practitioner.

3 Anti-ram Vehicle Barrier: The First Design Concept

In the early 2004, the government of the Republic of Singapore announced a tender for the development of a mobile crash barrier. Vehicle barriers are often used to stop unauthorised vehicles from entering protected areas. The Singapore government intended to deploy a crash barrier to enhance security for the 117th International Olympic Committee Session, which was scheduled to meet in Singapore in July 2005. In accordance with the tender specifications, the mobile crash barrier was expected to be capable of stopping a vehicle weighing up to 5 tonnes travelling at 50 km/h within 15 m. The tender requested that the barrier be able to be towed by a car with the kerb weight under 2 tonnes all over Singapore, be easily manoeuvred on site without a crane and be set up within 15 min at any location and without surface preparation. Other stringent crash barrier requirements included extended operational usage (up to 600 cycles in 9 h and up to 200 cycles an hour) while powered by a standard 12 V DC battery alone, fast response time of less than 4 s, as well as swift emergency activation of less than 1 s.

Although engineers of ST Kinetics have never previously designed nor built crash barriers, the company management was determined to bid for the tender. A small team consisting of four company design engineers, together with the employees of a potential manufacturing subcontractor, began a feasibility study on the design of the mobile crash barrier in February 2004. Before conceptualising their design proposal, the team analysed tender specifications, studied existing crash barrier designs and conducted a market survey. This research identified at least 15 suppliers of anti-ram vehicle barriers, which were certified by the US Department of State. Most of the identified crash barriers required permanent installation. Only two companies offered solutions for portable surface-mounted barriers that could be deployed quickly and did not require surface preparation.

It took the team around 4 weeks to propose a structure for the crash barrier that was robust enough to stop a heavy vehicle within the required 15 m (finite element analysis was utilised to ensure that all the structural elements were capable of withstanding the crashing load from the vehicle). This original crash barrier design is shown in Fig. 1.

The design presented in Fig. 1 consisted of a blocking plate *1* hinged to the body of crash barrier *2* and positioned between two box-shaped structures *3* that were coupled with an additional 5-tonne ballasting weight (not shown). The blocking

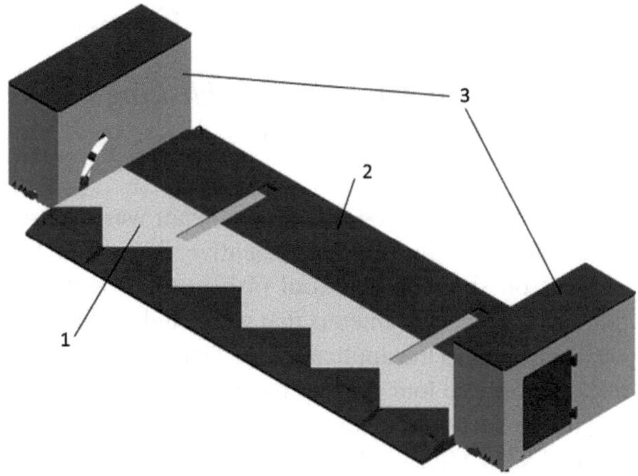

Fig. 1 The original design of a mobile crash barrier

plate *1* could be raised (to stop a vehicle) or lowered (to permit a vehicle to pass) by a hydraulic mechanism, located inside the box-shaped structures *3*.

Although the proposed design was sufficient to stop a heavy vehicle within the specified distance, the team was uncertain of its ability to win the tender. Firstly, the design did not meet all the stringent specifications of the tender. The barrier with the ballasting weight was too heavy to be towed by a car with the kerb weight of 2 tonnes. The time required to set it up significantly exceeded one specified in the tender. The team was also unsure how to deliver the required electric power capability and swift emergency activation. Secondly, the design did not look innovative enough to win. It was very similar to some existing designs identified by the team and was unlikely to defeat the proposals of other competitors. It was expected that a number of the international competitors, including the market leaders, would also be submitting tenders. These market leaders had significant advantage over ST Kinetics—their track record in developing crash barriers was extensive. Thirdly, the team was concerned about the competition from local companies. If a local competitor with much lesser production costs proposed a similar design, it would outperform ST Kinetics on the tender price. This latter possibility was anticipated very strongly—at the time when the first design was finalised, the subcontractor that originally worked with ST Kinetics left the team and decided to prepare a bid independently.

4 Applying TRIZ Tools: The Creative Design

4.1 Situation Analysis: Establishing the Needs and the Design Focus

In the mid-June of 2004, the project team was engaged in four TRIZ sessions that were conducted during 2 consecutive days. The first day was devoted to two sessions on situation appraisal (4 h in total) and resulted in the identification of four crash barrier parameters that were vital for the designs success. The second day was dedicated to the application of TRIZ tools to generate innovative design ideas relevant to these four priority parameters (3 h in total).

For the situation appraisal, the team used the Situation Analysis (SA) tool (see Appendix 1) (Belski 2009) and the product 'value curve' analysis (Kim and Mauborgne 1997). As a result, the team confirmed the requirements for the ST Kinetics' crash barrier which were essential to outperform the competitors and to win the tender. The following are the four priority parameters of the crash barrier that the team decided to focus at during the idea generation session:

Transportability Although tender specifications allowed a crash barrier weight of up to 7 tonnes and a size of up to 6 m × 3 m × 2 m, the team envisaged a much lighter and much smaller barrier. Small and light barrier could easily meet the towing requirements and could also minimise restrictions on roads that are suitable for its towing. The small size and light weight of the barrier would also improve its manoeuvrability (many roads in Singapore are narrow) and could eliminate the need for a crane to load and unload it.

Stopping Distance The team hoped to deliver a crash barrier with a stopping distance that was less than that specified under the tender requirements.

Power Requirements In order to raise/lower a blocking plate, existing crash barriers were either operated by AC power from the mains or by a 12 V DC battery. The latter was only used for emergency purposes and offered up to 40 operating cycles. This was significantly less than 600 cycles (200 cycles per hour) specified in the tender. The AC power use was not allowed by the tender. Therefore, the team needed to design a barrier with much lower power consumption than the existing models.

Emergency Activation The team believed that providing an opportunity to the crash barrier operator to activate the barrier urgently could significantly improve ST Kinetics' chance of being awarded the tender, because neither of the competitors were likely to offer fast barrier activation.

4.2 Generating Ideas: Applying TRIZ Tools

During the second day of TRIZ facilitation, the team concentrated its efforts on improvement of the above-mentioned four priority parameters. The team formally applied Method of the Ideal Result (MIR) (Belski 1998) and also used the 40 Innovative Principles (Altshuller 1984) and the Principles of Separation (Fey and Rivin 2005). While following the MIR procedure, the team (1) formulated the Ideal Ultimate Result (IUR) that covered all the four priority parameters, (2) explored the natural effects that helped/prevented achieving the IUR and (3) formulated a set of Target Tasks (TT) to accomplish. The team also catalogued all the available resources that can be freely used to implement the Target Tasks (see Appendix 2 for the list of the identified resources). The following is a simplified record of the solution process.

The IUR for the crash barrier that incorporated all the four priority parameters had been formulated as: *the ideal crash barrier is infinitely small and weightless, is able to stop an infinitely heavy vehicle, operates on its own internal energy alone and is always ready to stop an intruding vehicle.*

After evaluating the natural effects that prohibited achievement of the IUR (e.g. *Why an infinitely small and weightless barrier cannot stop an infinitely heave vehicle?)* and can help in fulfilling the IUR (e.g. *What existing concepts and natural effects may help to stop a heavy vehicle?)*, the following three Target Tasks were formulated:

- TT1: crash barrier needs to transform from lightweight while transported to heavy when installed on the road.
- TT2: crash barrier needs to generate sufficient power to operate (raise and lower the blocking plate)
- TT3: crash barrier needs to be swift in activation in case of emergency

Each TT was considered by the team separately.

4.2.1 Target Task 1: Lightweight Becomes Heavy

While performing the next step of MIR and analysing the resources that could help in transforming the crash barrier from lightweight to heavy, the team singled out the resource of the *road* (section of *substance resources* in Appendix 2). It suddenly became evident to the team that if it were possible to firmly affix the crash barrier to the road, even a very light barrier would become infinitely heavy. The team liked this idea and considered various options of affixing the barrier to the road.

The concept of making the crash barrier with many spears that could be inserted into the road to firmly attach it was the most appealing. While thinking of 'gluing' the barrier to the road, the team realised that the TRIZ principle of *Separation in Time* suited the situation well. Firm attachment of the crash barrier to the road was only necessary to stop the oncoming vehicle. It was not required under any other

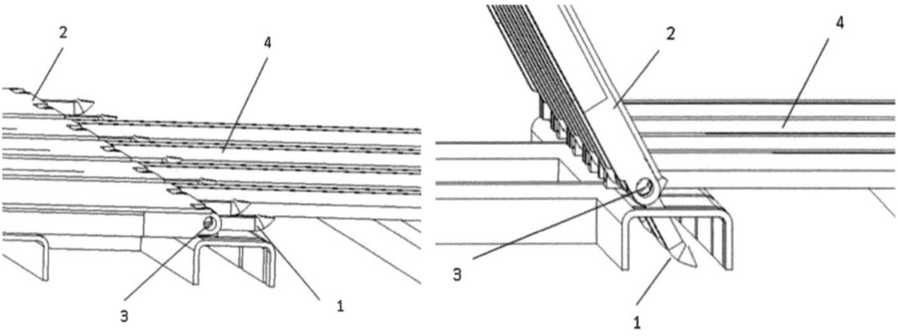

Fig. 2 The blocking (spears) frame in passing (*left*) and stopping (*right*) positions

circumstances. Thus, some powerful and rapid means to force the spears into the road and to firmly affix the barrier to the road at the time when the oncoming vehicle approached the barrier were required.

Further analysis of resources established that another substance resource—the *impacting vehicle* itself—could help in binding the barrier to the road. One way to realise such a design was to make the blocking plate from many spears. This 'blocking frame' can be positioned horizontally, close to ground level, in order to permit vehicles to pass or could be raised to prevent entry. If a driver decided to drive over the barrier when the spears' blocking frame was raised, the vehicle would force the spears to dig deeply into the road. This would attach the barrier to the road and 'make' it much heavier. The proposed structure of the blocking frame is shown in Fig. 2 with the spears' blocking frame in passing (horizontal) and stopping (angle) positions.

As shown in Fig. 2, an array of pointed teeth *1* is incorporated into the blocking frame *2* that is attached to the base frame *4* to allow rotation along the axis *3*. When the blocking frame *2* is fully lowered (left), the array of pointed teeth *1* flush with the base frame *4*. When the blocking frame *2* is in a stopping position (right) and is impacted by a vehicle, the force exerted by the vehicle on the blocking frame *2* levers the array of pointed teeth *1* into the ground.

The idea of a crash barrier that is affixed to the road underneath and made heavy by the intruding vehicle itself was very appealing. The design concept began to look very innovative. The team was excited and continued to scrutinise other primary parameters with a renewed wave of enthusiasm.

4.2.2 Target Task 2: Generating Power

The blocking frame needed the ability to be raised and/or lowered at any time in order to permit approved vehicles to pass and to prevent unauthorised entry. Could the barrier itself generate energy for raising and lowering the spears? Analysing the resources (Appendix 2) that could enable the crash barrier to generate energy for its

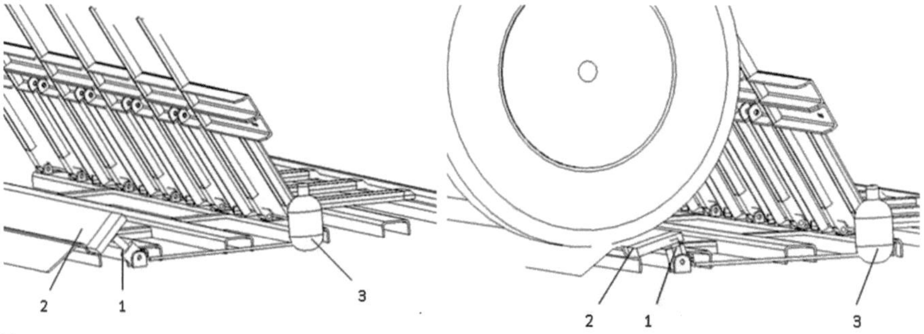

Fig. 3 Deployment a hydraulic accumulator to collect energy from the passing vehicles: (*left*) ramp released, (*right*) ramp pressed

own operations helped the team to identify another substance resource of the *vehicles permitted to pass* which appeared useful in achieving the second Target Task. When the crash barrier is positioned on a road, numerous vehicles that are permitted to pass through drive across it. These passing vehicles can be ‘requested’ to ‘recharge’ the barrier. The idea of using a spring mechanism to store energy was proposed first. The team then analysed power resources that had been identified (Appendix 2) and established two existing resources: *12 V DC battery* and *hydraulic power*. The team member who was an expert in hydraulics convinced the team that a hydraulic accumulator was easier to implement than a DC accumulator and also that it would be ideal for collecting energy from the passing vehicles. Figure 3 shows the implementation of this idea.

The ingress ramp *2* in Fig. 3 is connected to the hydraulic cylinder *1* which, in turn, is connected to the hydraulic accumulator *3*. Any passing vehicle depresses the ramp *2* and compresses the hydraulic cylinder *1*, forcing hydraulic fluid (not shown) into the accumulator *3*. As a result, a small amount of energy is stored by the accumulator every time a vehicle drives across the barrier. The more vehicles passing the crash barrier, the more energy is stored by the hydraulic accumulator to operate the crash barrier.

4.2.3 Target Task 3: Fast Emergency Activation

While the team was considering the third TT, it identified two energy resources that looked promising for delivery of fast emergency activation: *hydraulic power* and *electric power*, as well as the field resource of oil *compression*. The team was quick to establish a way to utilise those resources in an emergency situation. The side elevation of the crash barrier design that had been conceptualised after considerations of TT1 and TT2 is shown in Fig. 4.

Under the normal mode of operations, the blocking frame *1* in Fig. 4 is raised and lowered by means of a pair of hydraulic cylinders *2*. The latter are attached to the main frame of the crash barrier *3* on one side and to the blocking frame *1* on the

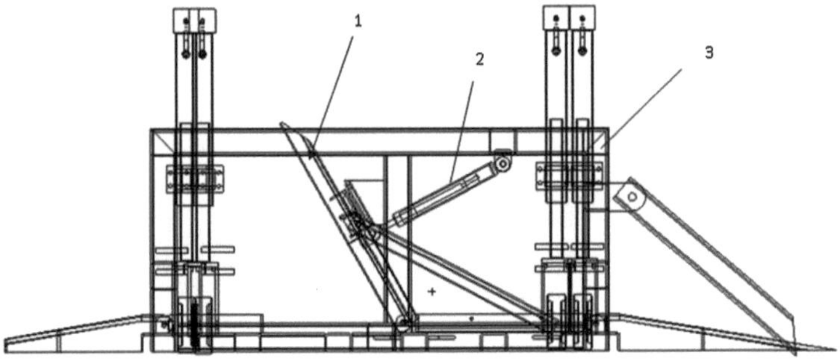

Fig. 4 Side elevation of the mobile crash barrier

other. The hydraulic cylinders *2* are driven by the fluid supplied by the flow-regulated valve (not shown). The valve reduces the pressure of the hydraulic fluid that is supplied by the hydraulic accumulator (not shown) to ensure safety of operations. Under the circumstances, therefore, the time required to raise/lower the blocking frame is from 3 to 4 s.

The team realised that the hydraulic accumulator possessed a sufficient resource of the hydraulic fluid pressure (up to 100 bar) that could be used for emergency activation. The team decided that in the emergency mode, the flow-regulated valve could be bypassed. The hydraulic cylinders that were supplied directly from the hydraulic accumulator could act much quicker. This would reduce the safety of operation somewhat, but could bring down the time of raising the blocking frame significantly. Implementation of the emergency activation was successful. The time to raise the blocking frame was reduced to just 1 s.

4.2.4 40 Innovative Principles: Improving Transportability

The application of MIR assisted the team to establish the key design ideas of the mobile crash barrier. It was necessary to put these ideas together and to ensure easy transportability and manoeuvrability of the crash barrier. In order to achieve these and to identify some creative analogies, the team scrutinised the list of 40 Innovative Principles of TRIZ (Altshuller 1984). The analogies proposed by the following principles helped the team members to put all the above-mentioned ideas together and to significantly improve transportability:

Segmentation: *a. If an object is uniform, make it modular. b. Make an object demountable. c. If an object is already modular, subdivide it further.*
Detachment: *a. Detach (remove or separate) the disturbing part or property from an object. b. Detach (remove or separate) the essential part or property from an object.*
Combination: *a. Combine like objects or objects intended for similar or related operations. b. Carry out similar or related operations simultaneously.*

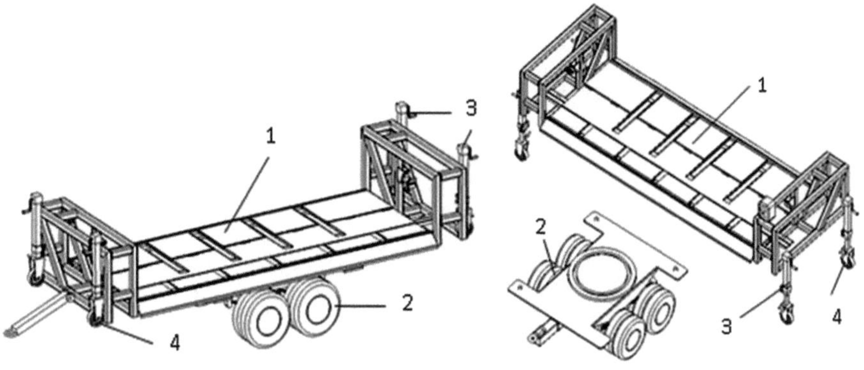

Fig. 5 Transporting a crash barrier: towing

As a result, the team proposed a design that offered two different options for transporting the vehicle barrier: by towing it and by moving it in the folded state. The option of towing the barrier is presented in Fig. 5.

During transportation, the main body of the crash barrier *1* is mounted onto the trailer system *2*. After the barrier is transported to the required location, the trailer system *2* is removed. This is done by means of the four hydraulic jacks *3* that are affixed to the main body of the barrier *1* and are raised during transportation. Firstly, these jacks *3* are extended to touch the ground. Then, the trailer *2* is detached from the barrier *1* and the length of the hydraulic jacks *3* is extended further to lift the barrier *1* off the trailer *2*. After the trailer *2* is moved away, the hydraulic jacks *3* are retracted to allow the main body of the crash barrier *1* to rest on the ground. It needs to be noted that the small wheels *4* attached to the hydraulic jacks *3* allow for small adjustments of the crash barrier's position.

Figure 6 depicts the vehicle barrier in a folded state.

The side frames *1* are mounted to the base frame *2* to allow frame detachment and placement on top of the platform elements. Alternatively, the side frames *1* may be hinged to the base frame *2* so that they can be folded onto the base frame *2*. The ingress ramp *3* and egress ramp *4* are made of four main parts each. To minimise the size of the crash barrier, the ingress ramp *3* and egress ramp *4* are hinged to the base frame *2* to allow folding the ramps perpendicular to the base frame *2*. As a result of this flexible design, the overall length of the folded vehicle barrier was reduced further. This significantly lowered the turning radius during transportation.

4.2.5 Novel Crash Barrier: Overall Implementation

After the TRIZ facilitation sessions, the project team combined all the design concepts that had been presented so far and finalised the design of the mobile crash barrier. The weight of the crash barrier was reduced to 2.5 tonnes. This ensured that an ordinary four-wheel drive could tow the barrier. Its size was reduced from the original 6 m × 3 m × 2 m to 4.2 m × 3 m × 2 m. The final design installed on the road is presented in Fig. 7 (Tay et al. 2006).

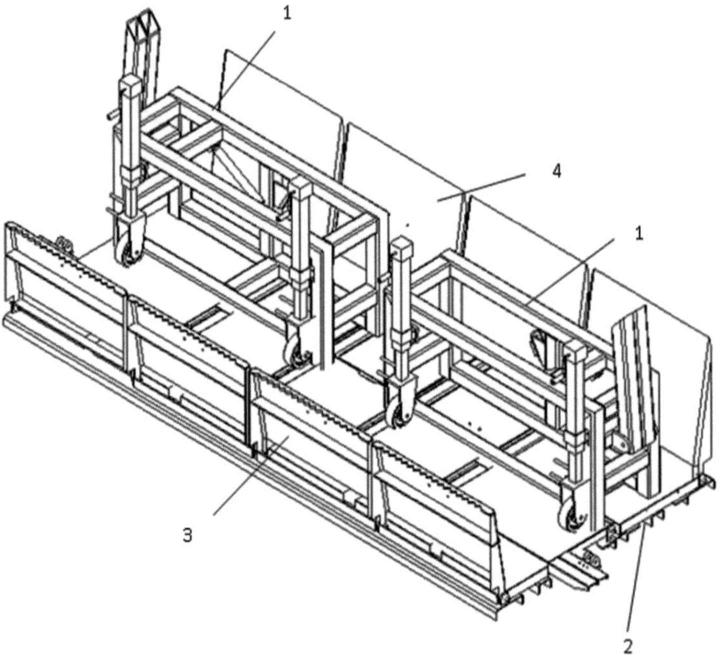

Fig. 6 Transporting a crash barrier: folded

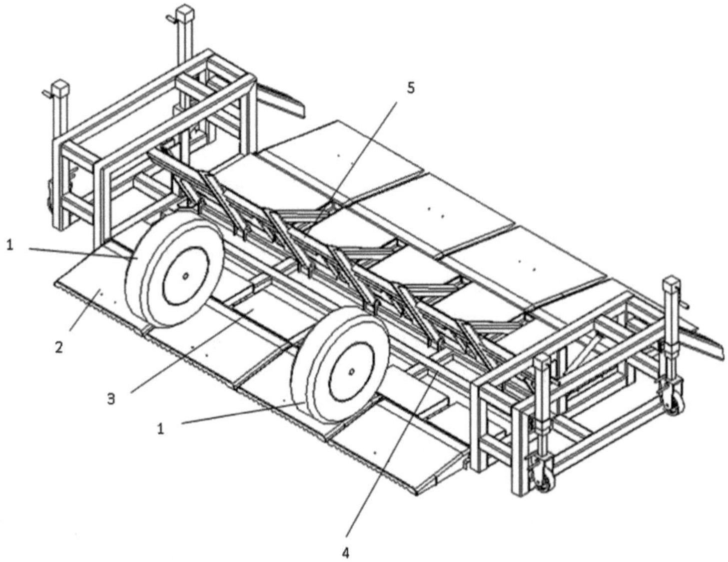

Fig. 7 Final design of the mobile crash barrier

Figure 7 shows a pair of the vehicle wheels *1* resting on part of the ingress ramp *2* and the platform elements *3* on the base frame *4*, when the vehicle is refused entry by the blocking frame *5*.

The team's design was successfully implemented, tested and patented (Tay et al. 2006). ST Kinetics won the tender and supplied the Singapore government with the mobile crash barriers that were deployed during the International Olympic Committee meeting in 2005.

5 Discussion

The original design of the crash barrier was ordinary. Similarly to many existing designs, it utilised a massive metal body that was heavy enough to stop a vehicle within the required distance under the conditions of direct vehicle-barrier collision. The second design of the crash barrier that was produced during the TRIZ facilitation sessions was elegant and won the tender, was patented, was deployed during the International Olympic Committee meeting and was even praised by Lee Kuan Yew.

This substantial difference in novelty of the two designs could only be explained by the engagement of the project team in TRIZ facilitation sessions. There are actually two possible explanations for the novelty of the second design. Firstly, the tools of TRIZ could have infused the creative strategies onto the project team. Secondly, it is possible that the creative design ideas originally arose in the head of the TRIZ facilitator himself and were passed onto the team during the facilitation sessions. The latter explanation can be ruled out immediately—the facilitator did not suggest any specific design ideas to the team. He only ensured that at the sessions the team members were using the TRIZ tools correctly. Assuming that the former explanation of novelty is the right one, let us explore how the TRIZ tools infused creative strategies onto the designers.

The diary entries reveal that the situation appraisal session influenced the project team profoundly. Firstly, the SA tool aligned the team and prepared the team members to the idea generation process by focusing them all in a unified direction. Secondly, SA also infused the strategy of extensive analysis and reframing (strategy i) onto the team. The questions of the SA tool, as well as the establishment of the 'value curve', engaged the team in at least four problem-framing activities that have been used by the creative designers. All 14 questions of the SA tool engaged the team in re-evaluating the barrier's application (activity c) and in studying the problem situation in depth (activity e). Questions 7–9 specifically guided the team in clarifying the relationships between the crash barrier and its supersystem (activity f). Questions 12 and 13 of the SA tool, as well as the 'value curve' development, further engaged the team in identifying the key issues to be addressed by the design (activity d).

MIR infused both the reframing strategy (strategy i) and the strategy of designing from the first principle (strategy ii). The former was achieved by engaging the team members in reassessing the problem from the first principles (activity b). This was facilitated by formulating the IUR, by reformulating it into TTs and by listing the system's resources. A well-formulated IUR is never achievable (e.g. *the ideal crash barrier is infinitely small and weightless, is able to stop an infinitely heavy vehicle*). Therefore, when users state the IUR, they are usually puzzled for a little while because the IUR sounds very unusual (e.g. *how could a weightless barrier stop an infinitely heavy vehicle?*). The TT is achievable and realistic. In order to state the TT, a practitioner investigates the reasons that preclude the IUR from occurring based on the principles of physics. These activities can trigger the reframing of the original task and significantly change the practitioner's perception of the problematic situation.

Formulating the IUR and restating it as TTs helped the crash barrier team to see their task differently. For example, the first part of the IUR formulated by the team suggested that *the ideal crash barrier is infinitely small and weightless and is able to stop an infinitely heavy vehicle*. In order to restate this IUR as a suitable TT, the team had to establish why and how the laws of physics do not allow a light barrier to stop a heavy vehicle (answer the questions: *What natural effects oppose the achievement of the IUR?* and *How do natural effects oppose the achievement of the IUR?*). Later on, the team had to search for the resources (Appendix 2) and for the ideas to implement the TT (*What existing concepts and natural effects may be helpful in delivering the TTs?*). These activities assisted the team in reframing the original situation.

It can be argued that similar activities had already been performed by the team while working on the original design, so that the above-mentioned problem framing had taken place before the IUR was formulated. In order to establish the size and the weight of the structure capable of stopping a vehicle of 5 tonnes travelling at 50 km/h within 15 m, the team had already utilised the basic principles of mechanics and the laws of conservation of momentum and energy. How did the MIR-activated evaluation of physics principles differ from the assessment of the laws of physics during the original design? When the original design was developed, the laws of physics were specifically used to establish a weight of the crash barrier that was sufficient to stop a vehicle as per tender specifications. On the other hand, when the IUR of the barrier was formulated, the team was 'compelled' to see the problem from a different angle and to revisit the physics principles behind a simple collision. The MIR procedure required the team to think of ways to stop a heavy vehicle by a light crash barrier (IUR) and to research the means for making the light barrier heavy (TT1). The traditional viewpoint of a simple barrier-vehicle collision was challenged. This triggered the reframing of the original task.

It is interesting to note that both designs attained sufficient crash barrier weight in a similar way—by attaching an additional ballasting weigh to the main body of the barrier. The original design deployed the 5-tonnes ballasting weight that was connected to the main body. This ballasting weight had to be transported with the crash barrier. The second design utilised the 'ballasting weight' of the Earth instead. This 'ballasting weight' did not require transportation. Clearly, a small change of viewpoint (from *What weight of the barrier is required to stop a vehicle?* to *How can a light barrier stop a heavy vehicle?*) resulted in a substantial creative design leap.

MIR also engaged the users in designing from the first principle (strategy ii). This strategy was infused by searching for the ideas to implement the TTs (*What existing concepts and natural effects may be helpful in delivering the TTs?*) and by sifting through the resources (Appendix 2) that can help in achieving the TTs. For example, when the team considered the TT2 (*crash barrier needs to generate sufficient power to operate*), the MIR procedure guided the team in establishing the sources of 'free' energy that were associated with the barrier deployment conditions (*vehicles permitted to pass*) as well as the ways and the means to collect and to store this 'free' energy (*hydraulic accumulation*). Similarly, the TT3 (*crash barrier needs to be activated fast in case of emergency*) was delivered from the first principles: by identifying the existing energy resources and establishing how these resources could be deployed to achieve emergency activation.

As previously reported, the 40 Innovative Principles represent a set of hints which are derived from thousands of successful technical solutions (Belski and Belski 2008). They offer cues which relate to the principle's name and can help an experienced practitioner to quickly establish analogies that can be useful in solving new problems. The crash barrier team had used only three principles (segmentation, detachment and combination). These three principles provided the team with enough analogies to not only improve transportability of the barrier but also to elegantly combine all the design ideas generated by the team during TRIZ facilitation sessions. As shown in Figs. 5 and 6, the crash barrier consisted of many modules (even the main body of the barrier was modularised). Some modules were detachable (e.g. trailer, ramps). Also, in order to make the barrier more adjustable, the hydraulic jacks (position *3* in Fig. 5) had small wheels (position *4* in Fig. 5)—a clear combination of the trailer on wheels and the main body on small wheels.

It must be noted that the investigation presented in this chapter has weaknesses. Like a majority of real development projects, it was not planned. Therefore, the activities that resulted in the first design of the crash barrier that is shown in Fig. 1 were not recorded and have been recollected by the participants. Also, due to company reasons, the documents related to this case study were approved for public release many years after development activities took place. As a result, the set of data that was available for analysis was substantially reduced—the authors did not survey or interview the team members.

The record of the TRIZ facilitation sessions, as well as the recollections of the facilitator and the project team members, indicates that the role of the TRIZ facilitator was confined to helping the team in utilising the TRIZ tools properly. Nonetheless, it is possible that his influence on the team was more substantial. Although the facilitator did not make any design suggestions explicitly, he may have unintentionally pushed the team towards the design ideas established by him alone.

The above-mentioned weaknesses can be minimised in future research which will be able to obtain more adequate and timely information on case studies that are performed under real design conditions.

6 Conclusion

An analysis of the TRIZ facilitation sessions on crash barrier design demonstrates that the TRIZ tools engaged the crash barrier design team in activities which resulted in the infusion of effective creative design strategies onto the team members. This confirms the hypothesis that TRIZ can help a practitioner to 'think like a creative designer'. It also suggests that many other ideation tools may be successful in engaging practitioners in activities that foster design creativity.

Appendix 1: The Situation Analysis Tool

Formulate a task as you see it:

1. What is the name of a **system** (product or process) under improvement? In case of the system to be new, state its generic name.
2. What is the specific useful function (SUF) of the **system**?
3. What is the main useful function (MUF) of the **system**?
4. List the main parameters of the **system.**
5. What main parameters of the **system** are to be improved?
6. What is the name of a *superior system*, the system up one step in hierarchy?
7. What needs of the supersystem require the **system** under consideration to be modified?
8. Can the needs of the supersystem be fulfilled in some other way, without modifying the **system** or by using some other existing system?
9. What human (customer) needs require the **system** (*superior system*) to be modified?
10. Can human (customer) needs be fulfilled in some other way without modifying the **system** (*superior system*) or by using some other existing system?
11. What is your main objective for the situation improvement?
12. Are there any ways to meet the objective without modifying the **system**?
13. Reformulate the problem if you see it differently.

Appendix 2: Resources Identified by the Project Team

Substance Resources Crash barrier, impacting vehicle, vehicles permitted to pass, road, operators, terrorist, other passengers

Energy Resources Electrical power from batteries or nearby building, kinetic energy from passing vehicles, kinetic energy from impacting vehicle, gravitation, hydraulic power, sun, wind, man power

Field (Information) Resources

- *Mechanical:* collision, friction, direct contact, compression, deformation, and others
- *Acoustic:* sound from vehicles and collision and others
- *Thermal:* heat from engine and others
- *Chemical:* petrol
- *Electric:* 12 V DC battery power
- *Magnetic*: might be applicable as crash barrier and vehicles are mostly steel structure
- *Biological:* operators, terrorist, other passengers, dog

Time Resources Retrieving crash barrier from storage, preparation for its towing operation, being towed on road, preparation for setting up crash barrier on site, setting up of crash barrier, operation of crash barrier, impacting vehicle approaching crash barrier, impacting vehicle hitting-on crash barrier, impacting vehicle stopped by crash barrier

Space Resources Road space in front and after the crash barrier, space immediately around the crash barrier, space available by individual crash barrier components

References

Akin, Ö. (1990). Necessary conditions for design expertise and creativity. *Design Studies, 11*(2), 107–113.

Altshuller, G. (1984). *Creativity as an exact science*. New York: Gordon & Breach.

Amabile, T. M. (1983). The social psychology of creativity: A componential conceptualization. *Journal of Personality and Social Psychology, 45*(2), 357–376.

Belski, I. (1998). I wish the work to be completed by itself, without my involvement: The method of the ideal result in engineering problem solving. In R. Chapman & R. Hunt (Eds.), *World innovation and strategy conference* (pp. 199–206). Sydney: University of Western Sydney.

Belski, I. (2009). Teaching thinking and problem solving at University: A course on TRIZ. *Creativity and Innovation Management, 18*(2), 101–108.

Belski, I., & Belski, I. (2008). Cognitive foundations of TRIZ problem-solving tools. In: Vaneker, T. (ed.) *Proceedings of the TRIZ-future conference* (pp. 95–102). University of Twente, The Netherlands.

Candy, L., & Edmonds, E. (1996). Creative design of the Lotus bicycle: Implications for knowledge support systems research. *Design Studies, 17*(1), 71–90.

Cross, N. (2001a). Achieving pleasure from purpose the methods of Kenneth Grange product designer. *The Design Journal, 4*(1), 48–58.

Cross, N. (2001b). Design cognition: Results from protocol and other empirical studies of design activity. In C. Eastman, W. M. McCracken, & W. C. Newstetter (Eds.), *Design knowing and learning: Cognition in design education* (pp. 79–103). Oxford: Elsevier.

Cross, N. (2003). The expertise of exceptional designers. In N. Cross & E. Edmonds (Eds.), *Expertise in design* (Vol. 6, pp. 23–35). Sydney: Creativity and Cognition Press, UTS.

Cross, N., & Cross, A. C. (1996). Winning by design: The methods of Gordon Murray, racing car designer. *Design Studies, 17*(1), 91–107.

Cross, N., & Cross, A. C. (1998). Expertise in engineering design. *Research in Engineering Design, 10*(3), 141–149.
Fey, V., & Rivin, E. I. (2005). *Innovation on demand.* Cambridge : Cambridge University Press.
Howard, T. J., Culley, S., & Dekoninck, E. (2008). Describing the creative design process by the integration of engineering design and cognitive psychology literature. *Design Studies, 29*(2), 160–180.
Kim, W. C., & Mauborgne, R. (1997). Value innovation: The strategic logic of high growth. *Harvard Business Review, 75*(1), 102.
Lawson, B. (1994). *Design in mind.* Oxford: Butterworth Architecture.
Lee, K. Y. (2005). Made-in-Singapore technology to fight terror. Accessed May 10, 2012, from http://www.mfa.gov.sg/content/dam/mfa/images/om/washington/newsletter/March05.pdf
Lewin, K. (1947). Frontiers in group dynamics II. Channels of group life; social planning and action research. *Human Relations, 1*(2), 143–153.
Maccoby, M. (1991). The innovative mind at work. *IEEE Spectrum, 28*(12), 23–35.
Mednick, S. (1962). The associative basis of the creative process. *Psychological Review, 69*(3), 220.
Roy, R. (1993). Case studies of creativity in innovative product development. *Design Studies, 14* (4), 423–443.
Shah, J. J., Smith, S. M., & Vargas-Hernandez, N. (2003). Metrics for measuring ideation effectiveness. *Design Studies, 24*(2), 111–134.
Simon, H. A. (1996). *The sciences of the artificial* (3rd ed.). Cambridge: MIT.
Tay, C. M., Teng, T. C., Sioh, E. L., Shi, A. J., & Sim, C. N. (2006). Vehicle barrier. WO 2006/088428.

TRIZ-Supported Development of an Allocation System for Sheet Metal Processing

A One-Day Case Study

Barbara Gronauer and Horst Th. Nähler

Abstract This case study presents the new development of an allocation system by using TRIZ tools. The allocation of sheet metal parts for further treatment is currently done manually. The demand for decreasing station times to utilize the processing machine to its full capacity brought up the need for a partially automated allocation system. Because of the high requirements regarding process reliability, safety, robustness, maintainability, and asset costs, a systematic approach using TRIZ tools was necessary for developing a successful solution concept. The concept definition was achieved during a 1-day workshop with developers with no prior training in TRIZ. The workshop was conducted by two TRIZ facilitators, who lead the developers through several stages of a TRIZ-based process: the first step covered the clarification of the task, the customer requirements, and specific company guidelines. Then analytical tools of TRIZ were used to structure the task and identify key problems and development obstacles. After that, TRIZ solution tools were utilized to develop ideas which were finally combined into a qualified, capable solution concept. The article shows how TRIZ tools sped up the process of task analysis, problem identification, and solution generation significantly.

Keywords TRIZ case study • Systematic product development • Automation • Allocation system • Sheet metal processing

1 Initial Situation

The case study describes the procedure and results of a 1-day workshop during an actual development project of a German Small Enterprise that is specialized in developing and building special purpose machinery. For this paper, the company

B. Gronauer (✉)
StrategieInnovation, Hünfeld, Germany
e-mail: bg@strategieinnovation.de

H.T. Nähler
c4pi—Center for Product-Innovation, Hünfeld, Germany
e-mail: naehler@c4pi.de

L. Chechurin (ed.), *Research and Practice on the Theory of Inventive Problem Solving (TRIZ)*, DOI 10.1007/978-3-319-31782-3_10

will be called GSE. Because of the actuality of the project and existing nondisclosure agreements, names, details, numbers, sketches, and drawings have been modified and generalized.

It was the company's target to speed up the development and solution finding process by using the TRIZ methodology, without having to train the team members in TRIZ. In the past, published case studies either described a process relying on trained workshop participants [exemplary described in Waldner and Dobrusskin (2014)] or the problems were tackled by single TRIZ specialists [exemplary described in Sire and Gillmann (2014), Benjaboonyazit (2014)]. Other case studies are presented with the purpose to teach TRIZ and have been modified to show the TRIZ methodology [e.g., Katolický et al. (2014); Gundlach and Nähler (2006)].

This article reflects a real-world case study, where a 1-day workshop was conducted by two trained facilitators (MATRIZ Level 3 certified and accredited). During the workshop the trainers guided the untrained participants through the single TRIZ tools. With appropriate preparation and language used, the TRIZ tools were easily understandable and usable even for untrained participants. The facilitators also used breaks during the workshop day to process input from the workshop participants and prepare the presentation, wording, and input for the next TRIZ tools.

The project was initiated by a customer of GSE, which is producing sheet metal parts of different sizes. During the manufacturing process, metal sheets are formed in presses. The formed raw parts then have to be processed in processing machines. Currently, one press supplies up to six processing machines with one worker supplying two processing machines. The processing takes up to 90 s which leaves enough time for the worker to manually place a raw part into the fixture of one processing machine and during processing to remove the finished parts and clean the fixture of the second machine, which is then fed with another raw part (see Fig. 1). In this setting, the long processing time in the processing machine is the bottleneck of the production flow.

To eliminate the bottleneck of the long processing time in the processing machine, the customer of GSE intends to significantly increase production output by replacing the processing machines with new ones which are capable of cutting the processing time in half. As a result, the worker at the processing machine becomes the bottleneck, and the whole process becomes more prone to human error and variations in productivity of the worker, according to calculations of GSE and its customer.

For that reason, the customer of GSE desires a (semi)automated solution which uncouples the allocation/supplying process from the processing itself. This should be achieved by providing a cache for raw parts which on one end is easy to load manually by a worker and which can on the other end be off-loaded by a robot capable of quickly and accurately placing the raw parts into the fixture of the processing machines as well as taking out the finished parts (see Fig. 2).

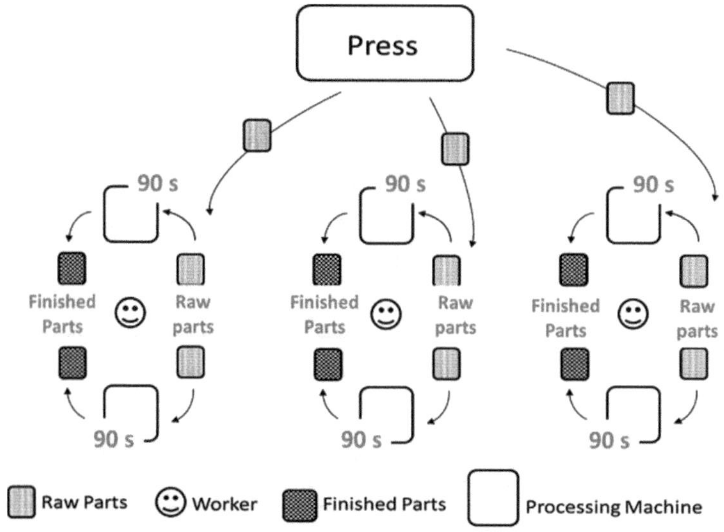

Fig. 1 Initial situation of the manufacturing process

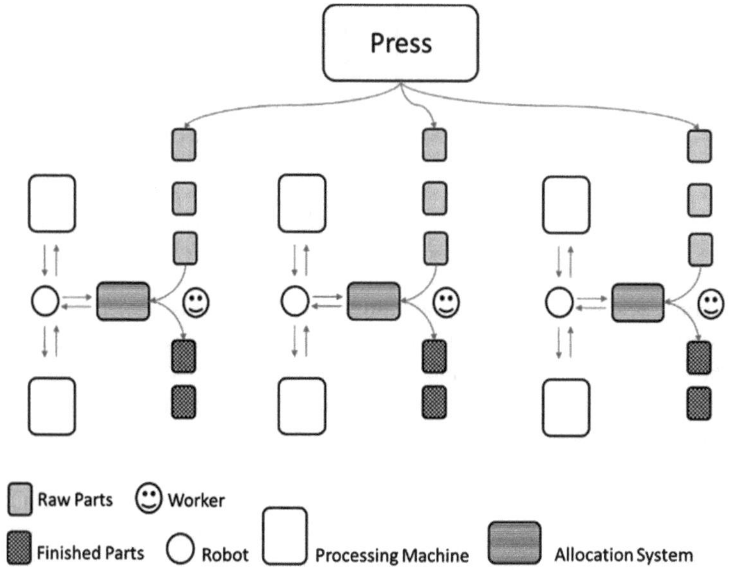

Fig. 2 Workflow with semiautomated allocation system as desired by the customer of GSE

1.1 Situation Clarification

TRIZ was introduced into the development process at a time where a rough concept for the allocation system was already at hand.

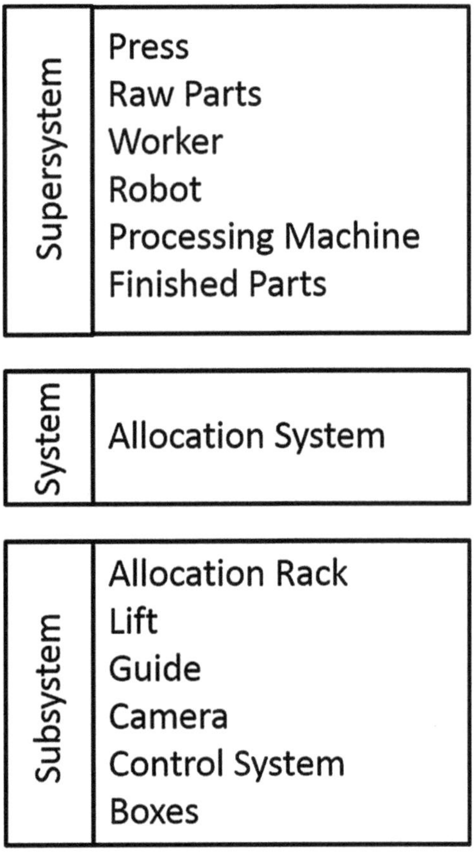

Fig. 3 Simplified representation of the rough first concept of the desired allocation system

The concept consists of an allocation rack which holds a specified number of transport boxes. These transport boxes contain raw parts and finished parts and allow the inflow and outflow to and from the processing machine on stacked levels. The boxes are placed on angled roller beams and are therefore able to move by themselves due to their mass and gravity.

On the loading end, the system features a guide that allows a controlled flow of the boxes to and from the rack to the point where the worker is loading/unloading the parts. The offloading end has a lift system that is bringing the boxes to the required floor of the rack and positions the boxes to allow the loading/offloading of parts by the robot. Furthermore, an image recognition system is used in conjunction with the robot to allow accurate gripping of the parts. A schematic sketch of the allocation system is shown in Fig. 3.

The most important customer requirement was, besides velocity and robustness, the safety of the worker. Nevertheless, details of most customer requirements were still unknown, fuzzy, and differing throughout subject matter experts and generally subject to change, so it was a typical business situation.

The first task was both to clarify the initial situation and known facts/requirements and to define how far reaching the solution should be. The TRIZ tool chosen was the multiscreen approach (Hentschel et al. 2010; Koltze and Souchkov 2011). With this approach, the team was able to look at the structure of the system to be developed, its environment (supersystems), and its current subassemblies (subsystems). Furthermore, by including the timeline (past-present-future), the team was enabled to systematically assess the reasons for the change from the "old" production process to the current or planned process as well as estimating changing conditions in the future which might affect current design decisions.

As a result of using the multiscreen scheme, the team was able to gather available information, generate substantiated estimations, and bring all this into a structured form in a short amount of time. Based on the scheme, decisions could be made affecting the design of the allocation system with regard to future requirements while making sure that current requirements are being met. The multiscreen approach worked as a catalyst in bringing together the knowledge of the team members and focusing their thoughts and estimations (Altshuller 2000).

1.1.1 First Step: System—Subsystems—Supersystems

In the first step of filling the nine screens, the team looked at the current stage of development (rough concept) and the intended design of the allocation system (see Fig. 4). The system is "allocation system," consisting of a guide, rack, lift, camera, control system, and transport boxes. The supersystems identified were the upstream press, raw parts, finished parts, worker, robot, and downstream processing machines.

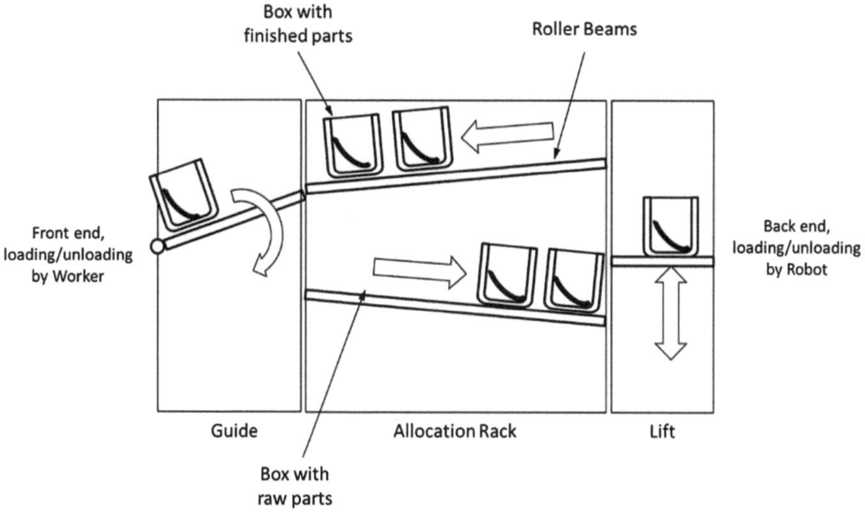

Fig. 4 System structure of the allocation system

1.1.2 Second Step: Past—Present—Future for Understanding the System

After structuring the current system design and raising awareness for the supersystem requirements and boundary conditions, the history of the system was assessed. In the past, the processing machines were loaded and unloaded manually by a worker. Other than that, the process worked similar. It is also recognizable that the system development follows the Trends of Engineering System Evolution with respect to decreasing human interaction (Mann 2013). In the past, the worker was the system delivering all the functions necessary from placing the raw part and removing the finished part (energy source, transmission, tool, control system (Altschuller 1998; Hentschel et al. 2010; Ivanov 2013), while in the current design, the worker is already partly disconnected from the process. The accurate placement of the raw parts is now more efficiently done by a robot and automated control system (see Fig. 5).

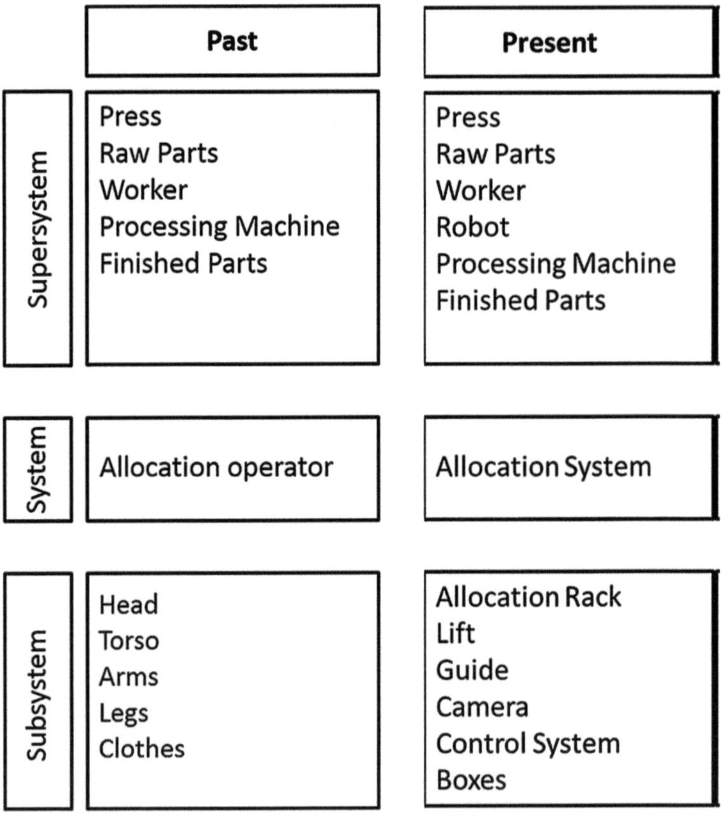

Fig. 5 Past and present system

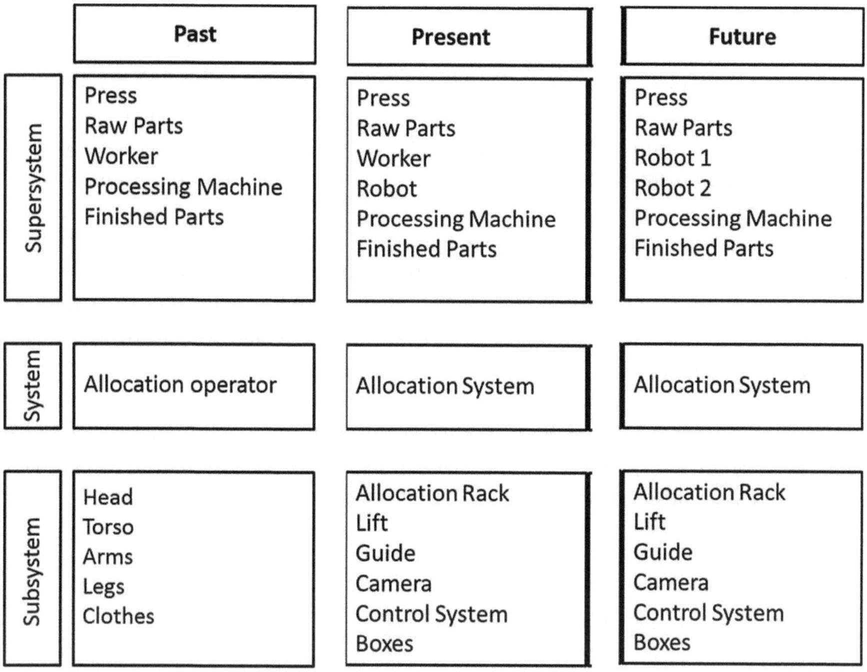

Fig. 6 Multiscreen assessment of the allocation system

While discussing about the future, it was clear that the worker will be replaced by another loading robot on the front end of the allocation system to create a fully automated process. During this discussion, the development team identified requirements resulting from that future scenario and how to detail the system today to be prepared best for future changes (see Fig. 6).

With this preparatory analysis done, the team could decide on which level the most significant tasks are and which problems and obstacles to tackle first. It could be made clear that the further detailing of the allocation system had top priority and that the interfaces between the robot and processing machines had to be addressed later. So the decision was made to use TRIZ to further analyze the allocation system and the tasks arising from the client's requirements.

2 Problem Analysis

Following the TRIZ process, a thorough problem analysis is necessary to identify main problems, build relevant problem models, and chose applicable solution models and solution strategies.

First, a function model of the allocation system was built. Even at this early stage, the components chosen as the general structure of the system could be used to sufficiently model the system with its interactions. During this process all team members agreed upon the terms used and the syntax of the TRIZ function analysis made it easy to communicate on a general level free of specialized expert's jargon. The first function model was aimed at representing a functioning system, free of disadvantages.

In the second step, function disadvantages have been added that represent current development obstacles and shortcomings of the current concept. By discussing these topics within the syntax of the function analysis, the problems could be objectified and the focus could be laid upon harmful, insufficient, and excessive interactions between the components of the function model.

After including the function disadvantages into the function model, each disadvantage was assessed, for some of them a cause and effect chain analysis (CECA) has been conducted. The assessment brought up several contradictions which have been addressed subsequently in the solution phase.

2.1 Function Model of the Allocation System

According to the rules of TRIZ function analysis (Hentschel et al. 2010; Training Material 2006), the main function of the allocation system was formulated as "Allocation System holds Parts." The interactions of the components have been formulated. Only the graphic representation of the function model is shown here (see Fig. 7); component analysis, interaction analysis, and the tabular function model are not explicitly shown.

Based on practical tests carried out with a test arrangement function disadvantages have been formulated to represent the problematic areas within the system (Fig. 8).

Four function disadvantages are exemplarily shown in the function model in Fig. 8.

The test showed that the boxes which hold the parts gather a significant speed while sliding on the roller beams. This high speed results in a sudden impact when the boxes need to be stopped. According to the requirements of the customer of GSE, the most important factor was the safety of the worker. Due to legal regulations, the system must under all circumstances provide safe and secure working conditions. The risk of injury due to moving parts has to be eliminated. Also, specified limits for noise emissions have to be met. These objectives could not be met with the initial rough concept of the allocation system:

1. Guide stops boxes excessively
   This disadvantage depicts the situation that the boxes, due to their weight and the angular positioning of the guide, have a high velocity that leads to an abrupt

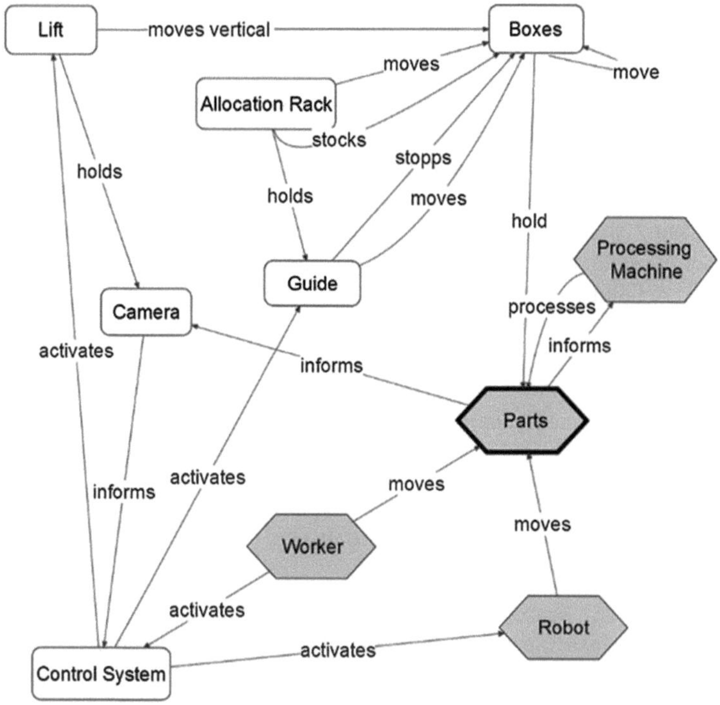

Fig. 7 Graphical function model of the allocation system

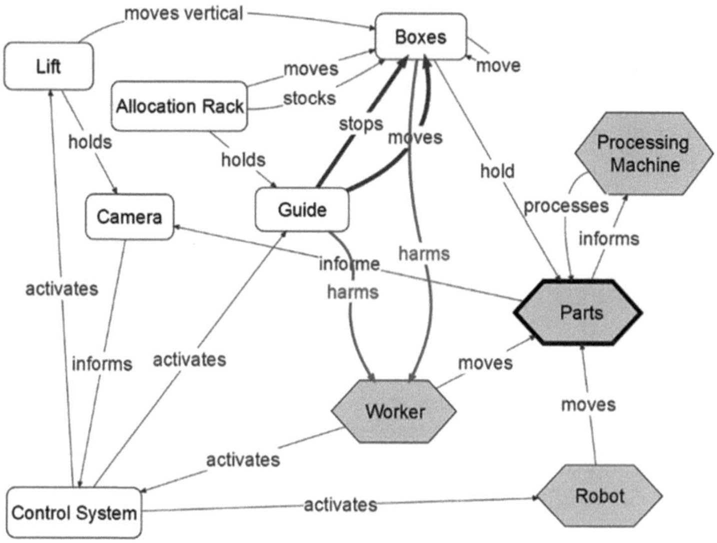

Fig. 8 Function model with function disadvantages

impact. This impact generates noise and represents a potential threat to the worker.

2. Guide harms worker
   The guide consists of moving parts to be able to direct the boxes into different levels of the allocation rack. These moving parts are positioned in the area where the worker has to load and unload parts. Therefore, the threat of injury is given and has to be eliminated.
3. Boxes harm worker
   As mentioned under 1, the boxes are able to hit the worker due to their kinetic energy. A hand can potentially be placed between parts of the guide and the moving box with the initial design concept.
4. Guide moves boxes excessively
   Due to the high angle of the guide, the boxes are accelerated to a high velocity, leading to high kinetic energy and thus being a risk for the worker.

2.2 Cause and Effect Chain Analysis

To uncover the key problems, the function disadvantages have been further analyzed using the cause and effect chain analysis. Starting with the apparent problem of the worker being harmed, cause and effect chains have been developed (Fig. 9).

The key problems are the circumstances that the "processing time is cut in half," "the length of the work space is limited," and "the boxes have to have a certain height."

Additionally, through the practical tests, the "operating zones" (Altshuller 2000) have been identified where injuries could happen. Exemplarily, these "operating zones" are at all edges where the sliding surfaces of the boxes change their angle relative to each other, e.g., the transition from the rack to the guide. Due to this change in angle, a gap is forming where a worker might place his hands and be injured (Fig. 10).

3 Formulating Problem Models

In the discussion around the causes and effects of each disadvantage, some ideas have been generated. Nevertheless, each idea was attached to subsequent disadvantages, so that engineering contradictions could be built starting with "initial ideas" that would lead to improvement in one aspect but also lead to deteriorating other factors. Some of the contradictions are listed below.

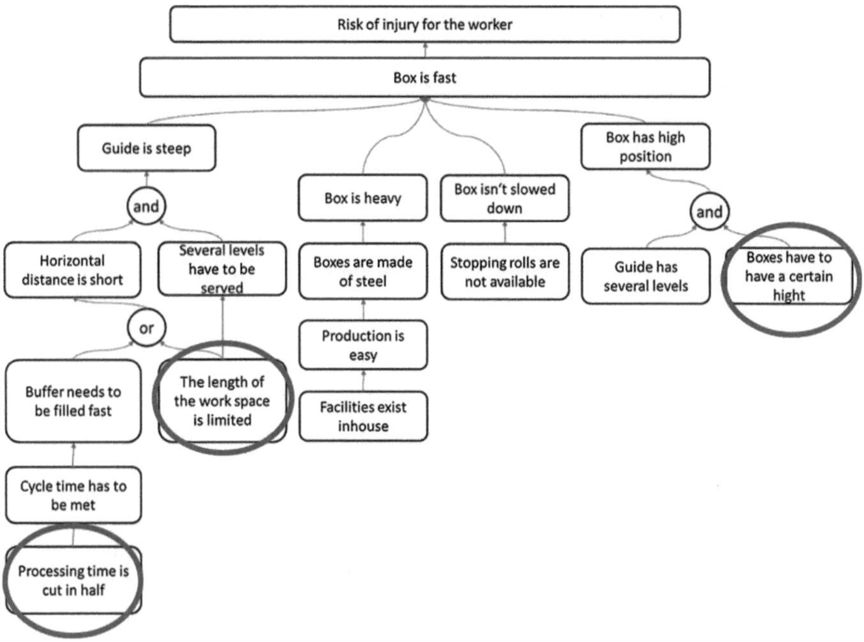

Fig. 9 Cause and effect chain analysis for risk of injury; circles mark key problems

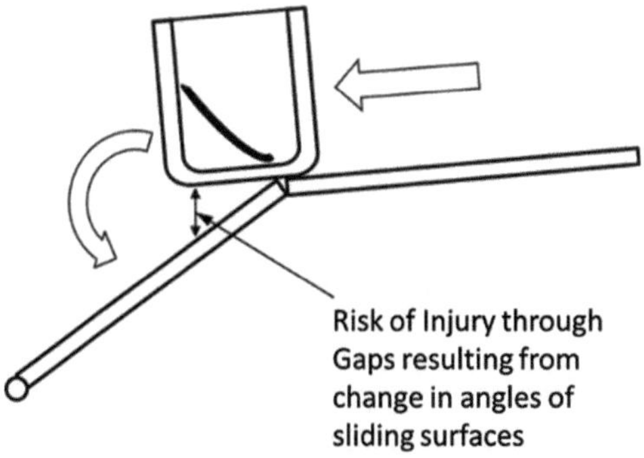

Fig. 10 "Operating zone" of initial rough concept where the problem occurs

3.1 *Primary Problem: Worker's Health and Safety*

According to the first design concept, the boxes containing the parts were designed to slide on roller beams through their own weight and the angular positioning of the guide and allocation rack levels. Preliminary tests showed that the boxes were really picking up speed along the length of the allocation rack. In the "operating zone," the boxes then rapidly tilt where the roller beams transition to the guide (see Fig. 10.). This tilting poses a risk for the worker as he might inadvertently place his hand in this area (see also CECA).

3.1.1 Engineering Contradictions Regarding Use of Steel Boxes

In the first attempt, the use of boxes was questioned and several options were discussed (e.g., belt conveyor with bristles to replace boxes and roller beams). However, using steel boxes as the means of transportation was preferred by the customer of GSE due to their in-house availability. Another advantage was the increased ability for the camera system to recognize size, shape, and orientation of the parts in the "defined" space of the box. Therefore, we concentrated on using boxes and stated their advantages and disadvantages in the form of engineering contradictions:

If	we use boxes
Then	a.) the allocation system is "simpler"
	b.) the parts have a "defined" space and are easy recognizable
But	a.) risk of injury is higher
	b.) weight is higher (boxes are heavy)

The formulation of engineering contradictions allows concentration of the parameters that we want to improve (stated in the "then…"-line) and its counterpart, the parameter that gets deteriorated as a consequence (stated in the "but…"--line). Converted into the language of the contradiction matrix [the 2010 version was used during the workshop Mann (2013)], the engineering contradiction then consisted of the conflicting parameters:

Allocation system is simpler	41—Manufacturability
	45—System complexity
Risk of injury is higher	37—Security
	38—Safety/vulnerability

All resulting inventive principles from the combination of the four parameters were then assessed:

		37		38	
		Worsening feature			
	Improving feature	Security		Safety/vulnerability	
41	Manufacturability	**1** **10**	**24** **13**	**6** **28**	**35** **9**
45	System complexity	**28** **24**	**5** **10**	**24** **26**	**10** **35**

The suggested inventive principles are:

1: Segmentation
24: Mediator
10: Preliminary action
13: "The other way around"

6: Universality
35: Parameter changes
28: Mechanics substitution
9: Preliminary anti-action
5: Merging
26: Copying

An excerpt of some ideas based on the inventive principles is:

1: Segmentation in combination with 24, mediator
Using a bristlelike structure applied to the guide to slow down the boxes and prevent the rapid tilting of the boxes

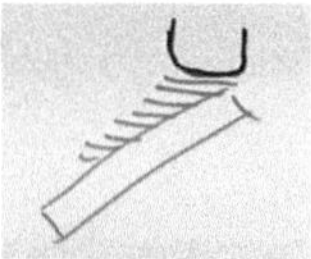

24: Mediator

Using a seesaw to tilt the boxes in a controlled way

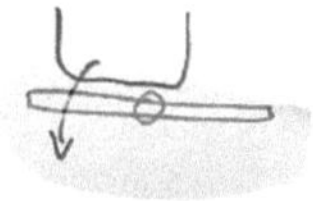

28: Mechanics substitution, suggestion b.) Use magnetic fields
Using a magnetic field to slow down the boxes during tilting

28: Mechanics substitution, suggestion c.) Use structured fields
Arrange the rollers of the roller beams in a x-type pattern (their rotation axes angled toward each other) to slow down the boxes

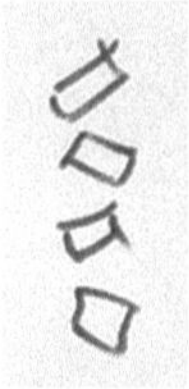

35: Parameter changes in combination with 5, merging
Use a curved transition to the guide to minimize the gap between the boxes and the roller beams

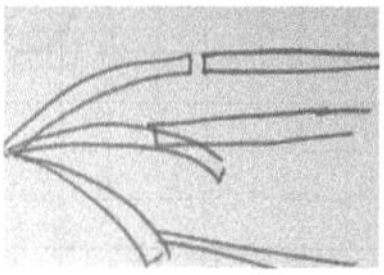

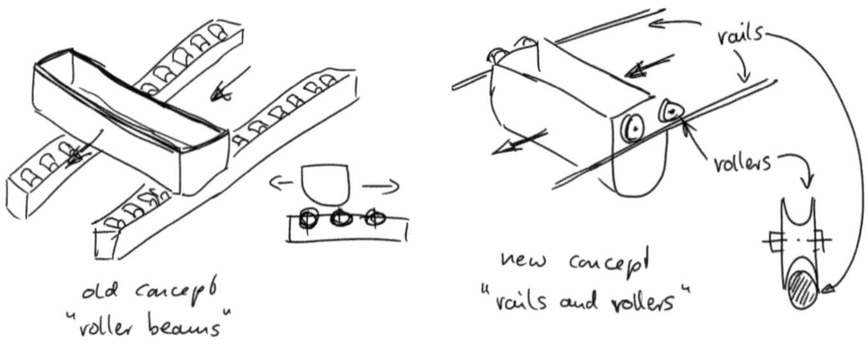

Fig. 11 Old concept, roller beams (*left*); new concept, rails and rollers (*right*)

13: The other way around
Putting the rollers on the boxes, not on the beams (little trolleys)

During the assessment of the ideas, the concept of using rollers on the boxes was the preferred concept. This was further substantiated, and to slow down the boxes on the guide, a counterweight was suggested that is pulled up by the trolleys

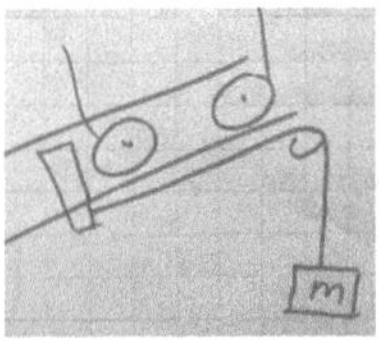

.

This concept was further changed to a roller-and-track system, where rollers are placed in the top corners of the boxes, which can then be hanged into a track system that is guiding the boxes throughout the allocation system (Fig. 11).

This concept also enabled the elimination of a driven belt on the back-end of the allocation system which before was necessary to move the boxes in and out of the lift. The roller-and-rail system allows the boxes to move themselves into and out of the lift. Small gaps and angular changes in the track can easily be rolled over.

The thought process was supported by the concept of Ideality, always looking for possibilities of functions being carried out by "itself" or looking for the ideal system that is performing the useful functions without harmful side effects (Hentschel et al. 2010; Nähler et al. 2012; Koltze and Souchkov 2011). As a consequence, the roller-and-track system then has been optimized to be self-cleaning by using shaped rollers that run on guiding rails.

3.1.2 Engineering Contradictions Regarding Velocity of Boxes

One problem had still to be solved. The use of a counterweight results in slower moving boxes. While this is good for the safety of the worker, it has a negative effect on the overall process speed, as the boxes take a longer time to travel the distance of the guide down to unloading. The contradiction to represent this situation was stated as follows:

If	we slow down the boxes
Then	the safety of the worker increases
But	the cycle time will be long

Instead of analyzing the engineering contradiction, this statement has been converted to a physical contradiction as follows.

3.1.3 Physical Contradiction Representing Worker Safety and Cycle Time

The following physical contradiction formulates the heart of the problem regarding the usage of steel boxes, its resulting risk of injury and the influence on cycle time. The formulation enables a focused idea generation around this inventive problem:

Physical contradiction:

The boxes should be moving fast to achieve a short cycle time.
They should be moving slowly to achieve a safe work environment.

The separation principle "separation in time" led us to an interesting solution concept:

By separating the need for fast moving boxes and still boxes in time, an enclosed solution was developed that only allows the worker to access the loading / unloading area when all moving parts have come to a full stop. The solution suggests a sliding door that is opened by a counterweight (ideally "opening itself") as soon as a box is ready to be loaded/unloaded. After the worker completed his step, he closes the sliding door, activating the allocation system and allowing the boxes to move fast through the steep guide and allocation rack. This results in lower cycle times while eliminating the risk of injury. Another advantage of an enclosed system is that noise can be effectively damped and no special dampers are necessary to reduce impact speed of the boxes (Fig. 12).

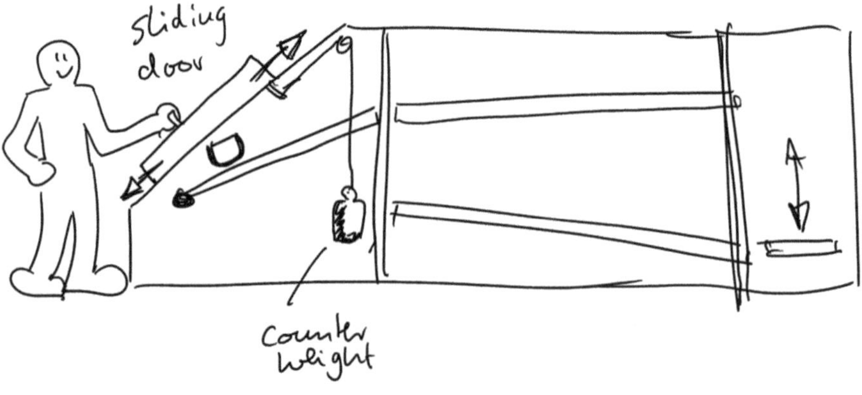

Fig. 12 Enclosed allocation system concept

4 Conclusions

The support of the development project with TRIZ resulted in a systematic approach to solving the technical problems at hand in a step-by-step approach.

The time invested for thoroughly analyzing the situation, identifying the major problems, and focusing on root causes instead of symptoms paid off regarding several aspects:

1. The TRIZ-guided process opened eyes for the "real" problems and in which order to solve them. Previous decisions based on assumptions could be revised and corrected, saving time and proactively considering hidden or unspoken customer needs.
2. Especially the CECA and function analysis helped to uncover missing or contradicting information from the customer, which could then be included in the specifications and requirements management.
3. TRIZ brought up the right questions through its ideality- and "ideal system"--focused approach. As a result, effective solutions were generated in a short amount of time. Consideration of the "self-principle" led to sound design decisions that were robust, safe, and cost-effective.

Through this way of procedure, the approach of the development task was very structured, a common language among the developers was established, the analysis of problem situations was enhanced, and the people were enabled to focus solving of the "right" problems instead of symptoms.

Since the workshop the concept has been detailed further and subsequent problems have been solved. Upcoming development tasks for future versions have already been identified, one of them being to decrease the overall length of the allocation system to reduce the occupied space on the shop floor.

With the chosen approach to guide untrained developers through a TRIZ workshop, the common preoccupation of "TRIZ is complex, time-consuming, and complicated" could be dissolved. The workshop participants were highly impressed by the results they achieved within one day. Moreover, it has been shown that even within a limited timeframe, a series of TRIZ tools aimed at problem clarification and problem selection, and problem solving could be deployed. Nevertheless, a thorough preparation and trained facilitators are necessary for an effective, goal-oriented process. Because of the efficiency of the TRIZ workshop, it is intended to establish TRIZ as an integral part of the development process of the company.

References

Altschuller, G. S. (1998). Erfinden Wege zur Lösung technischer Probleme, Hrsg. Prof. Dr. Martin G. Möhrle, Cottbus, 1998 PI.

Altshuller, G. S. (2000). The innovation algorithm translation by Lev Shulyak and Steven Rodman TIC.

Benjaboonyazit, T. (2014, September 4–6). Systematic approach to problem solving of low quality arc welding during pipeline maintenance using ARIZ. In: Kässi, T., & Souchkov, V. (Eds.), *Proceedings of the 10th International conference TRIZfest 2014*, Czech Association of Scientific and Technical Societies (CSVTS), Prague.

Gundlach, C., & Nähler, H. (2006). Innovation mit TRIZ Konzepte. Praxisanwendungen symposion Publishing, Werkzeuge.

Hentschel, C., Gundlach, C., & Nähler, H. (2010). TRIZ Innovation mit System Carl Hanser Verlag.

Ivanov, G. (2013). Formulas of creativity, Hrsg. Dr. Robert Adunka.

Katolický, Z., Bušov, B., & Bartlová, M. (2014, September 4–6). Small Turbojet engine innovation and TRIZ instruments. In: Kässi, T., & Souchkov V. (Eds.), *Proceedings of the 10th International conference TRIZfest 2014*, Czech Association of Scientific and Technical Societies (CSVTS), Prague.

Koltze, K., & Souchkov, V. (2011). Systematische Innovation Carls Hanser Verlag.

Mann, D. (2013). Matrix 2010 Re-Update der TRIZ Widerspruchsmatrix translated by Horst Nähler, Elisabeth Brilke, Hrsg. Horst Nähler, Barbara Gronauer, Carsten Gundlach.

Nähler, H., Reiffenscheid, C., & Schnittker, F. (2012). TRIZ IT! Der TRIZ Leitfaden Terex Corporation.

Sire, P., & Gillmann, D. (2014, October 29–31). TRIZ "knowledge spiral" (and first steps of ARIZ) solving Civil Engineering "thermal link" issue. In: *Proceedings of the TRIZ future conference 2014 Global innovation convention*, Ecole Polytechnique Fédérale de Lausanne, Lausanne.

Training Material TRIZ Level 3 Certification Training MATRIZ 2006 (2010).

Waldner, R., & Dobrusskin, C. (2014, October 29–31). Product improvement using TRIZ at Philips, a case study. In: *Proceedings of the TRIZ future conference 2014 Global innovation convention*, Ecole Polytechnique Fédérale de Lausanne, Lausanne.

TRIZ Events Increase Innovative Strength of Lean Product Development Processes

Christian M. Thurnes, Frank Zeihsel, Boris Zlotin, and Alla Zusman

Abstract New product development has to cope with different target dimensions. The need for a short time to market requires a very efficient process, and the need for innovation demands new and promising quality of the intellectual work that is done by all the people involved in the process. This work captures a very efficient process pattern out of the field of lean product development and enhances it by adding TRIZ processes for increased innovative power symbiotically (TRIZ or theory of inventive problem-solving). By this means, TRIZ concepts and approaches are utilized as lean events in an event-driven lean product development process.

Keywords Lean product development • Events • Systematic innovation • TRIZ • Innovation process • New product development process

1 Introduction: Integration of TRIZ into Lean Product Development

Lean product development is one of the most actual topics in the field of R&D management. More or less, lean product development is the attempt to copy the Japanese role model of a straight customer-oriented efficient product development. As almost all concepts are actually known as "lean," also lean product development is based on the production and management concepts used and to a great extend originated by Toyota. Starting in the 1970s of the last century, Toyota's concepts

C.M. Thurnes (✉)
University of Applied Sciences Kaiserslautern, Competence Center OPINNOMETH, Amerikastr. 1, 66482 Zweibrücken, Germany
e-mail: Christian.Thurnes@hs-kl.de

F. Zeihsel
Synnovating GmbH, Mozartstr. 25, 67655 Kaiserslautern, Germany
e-mail: zeihsel@synnovating.com

B. Zlotin • A. Zusman
Ideation International Inc., 32000 Northwestern Hyw., Farmington Hills, MI 48334, USA
e-mail: blzlotin@ideationtriz.com; azusman12@ideationtriz.com

L. Chechurin (ed.), *Research and Practice on the Theory of Inventive Problem Solving (TRIZ)*, DOI 10.1007/978-3-319-31782-3_11

like the Toyota Production System (TPS) got well known all over the world. Later also other practices and concepts of Toyota also regarding management, leadership, and product development followed, and many companies try to adapt these concepts to create their own "lean" strategy. But not every company is part of the automotive industries, and not every company has the size of the Toyota company, which is typically the template for all companies that are trying to adjust their processes toward "lean."

Besides the undoubted success of the originator of the lean principles, tools, and processes, the most interesting question for many other companies around the world is how to achieve a similar success under their own framework conditions. As a contribution to answer this question in part, this article gives an overview about TRIZ events as an expansion of the recently described TRIZ-driven innovation events (see Thurnes and Zeihsel 2014) that provide also small- and medium-sized enterprises with a good occasion, to enrich a high-efficient lean product development process with the innovative power of some TRIZ tools (TRIZ or theory of inventive problem-solving) to create an efficient and innovation-oriented product development process.

2 Practicable Lean Product Development: Event-Driven Process

In contrast to the triumphal march of lean production methods, the lean product development became known later and very slow. There may be several reasons. On the one hand, the resources needed to copy the original role model are not available for most of the companies. On the other hand, the variances between product development processes of different companies are higher than in production processes. And furthermore, how to balance creativity, innovation, and short process and lead times in a process framework for efficient product development remains an unanswerable question.

Regarding efficiency, Ron Mascitelli developed a "practical, event-driven process for maximizing speed, profits and quality" (Mascitelli 2011) in his book "Mastering Lean Product Development. Besides using elements from typical lean product development and custom development methodologies (see, e.g., Mascitelli 2006), he created a process structure that follows many lean principles and guidelines. But furthermore, this approach is best placed to be used by small- and medium-sized companies. It is event driven, and by number, type, and duration of events, the process is very customizable without losing the power of its efficient execution. The process description even covers very early stages of the process, so that more or less the framework could rather be considered as innovation process than as product development process (see Fig. 1).

The increased efficiency is generated by the combination of appropriate methods of lean product development and recent project management techniques. This

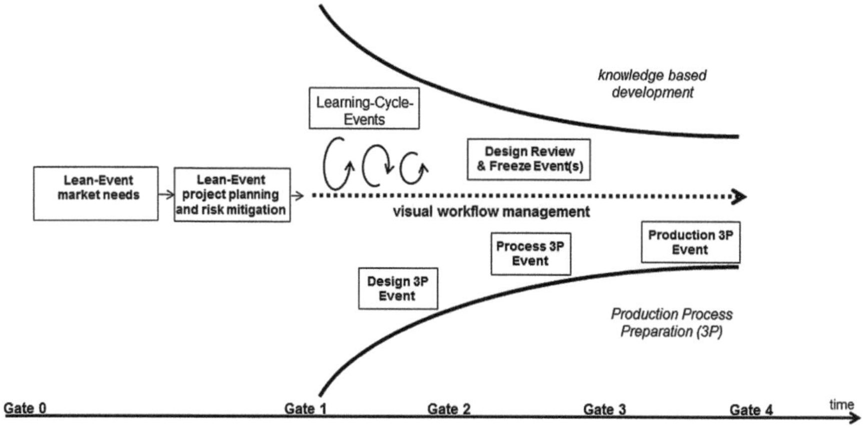

Fig. 1 Event-driven lean product development process according to (Mascitelli 2011)

allows independence from common phase-gate approaches. Small- and medium-sized companies often realize a better flow of activities, when their development process is controlled based on management by exception instead of management using rigid gates. However, if phase-gate is used, the gates may be integrated in the event-driven process.

3 Increasing Innovative Power Using TRIZ

The theory of inventive problem solving (TRIZ) is one of the most promising systematic methodologies, to incorporate innovation power in product development processes. Almost all of the TRIZ methods and tools used today are originated, influenced, or at least inspired by the works of Genrich S. Altshuller (see, e.g., Altshuller 1984, 2006; Altshuller et al. 1999) and his mentees.

The methodology disseminated from their Russian origin all over the world but in a smooth and gentle way. It evolved to a methodological toolset for many different innovation management tasks (e.g., see Terninko et al. 1998). Nowadays, some TRIZ specialists can be found in many companies and some organizations, like the European TRIZ Association (ETRIA) and the International TRIZ Association (MATRIZ), and others foster the usage and the ongoing research in this field. In Germany, a special group of the Verein Deutscher Ingenieure (VDI) founded an industry standard, which will be the fundament for a further dissemination of TRIZ in academics and practice (see Hiltmann et al. 2014).

Today the term TRIZ covers not only the fundamental works of Altshuller but also further developments and new methods. But most of the TRIZ tools and techniques can be used quite well in situations, when systematic procedures are needed, but on the other hand, a high degree of innovation in new solutions or products is aspired.

TRIZ nowadays is a set of many methods, tools, and principles. Even the classical TRIZ (see Altshuller et al. 1999) already was a bunch of many tools, and in this century, many additional approaches evolved. For the integration in the event-driven lean product development process, TRIZ is an appropriate tool to choose some tools, because they mostly are:

- Systematically structured
- Enabling creativity of many people with different qualifications
- Easily practicable (with help of experienced facilitators)
- Supporting a system-based approach
- Leading to specific solutions

Event-driven lean product development is predestinated for organizational learning and knowledge creation and concurrent efficiency. The integration of TRIZ increases the power of this process dramatically, because it adds systematic innovation to the already integrated systematic knowledge gathering. In other words, TRIZ adds an additional level of innovativeness to the process given.

4 Characteristics of Lean Events

As shown in Fig. 1, the concept of the event-driven lean product development is based on specific events. "Event" in this context is a specific working form, often used in lean approaches. The most known application might be the so-called Kaizen event, but also 5S events, SMED events, and other lean events exist—especially in the field of product development 3Ps' events are common (see, e.g., Bicheno and Holweg 2008 or Ward 2007). "3Ps" thereby stands for production, preparation, and process or similar expressions.

Events are scheduled in project plans but are more flexible as usual gates. They serve as points of concentrated teamwork to create important partial results and to ensure and foster an efficient flow in the development process. From an innovation perspective, each event creates new knowledge about the product to be developed. Especially in the lean product development at some points, even more knowledge is created that may be used in later projects. However, the event differs from usual gate meetings or project coordination meetings substantially. A lean event is not just a meeting, but it takes much more as an action. The most important factor is that the lean events create work results and not just decisions or action item lists (without really working on these items before). And these work results enable the flow of work till the next event. Any lean event as part of an event-driven process consists of (see Mascitelli 2011):

- A preparation phase prior to the event (starting 3–4 weeks before the actual event)

- The actual event (in our case mostly 1 day)
- A follow-up phase, while the tasks necessary to reach the next preparation phase should be fulfilled

An important factor for successful lean events is to have a specialized structured approach for every kind of lean event.

So the preparation phase encompasses usually an intensive data gathering. But this action happens in a structured way—determined by the kind of event. Also the definition of goals and limitations is part of the preparation phase as well as the definition of team members and resources in a standardized way.

The actual event does not have to have any timely exact schedule, but the activities have to be structured and well known by the participants or leaded by experienced facilitators. In this structured teamwork session, team members apply collaboratively standardized tools and methods along the standardized approach. At the end of the event, the results are documented in a standardized way, and the team develops the tasks that have to be fulfilled before the next event can be started. The structured form to depict the action items should be standardized.

Also the follow-up phase shows some structures. Action items are tracked and routinely assessed.

A key for the success of lean events is the high level of activity of the participants. These activity is the result of structured and standardized procedures in combination with motivating framing conditions and an experienced facilitation. It has to be considered that a big part of the motivation results in the strong emphasis on *doing* things and the will to create results in short time. For more detailed information about the characteristics of lean events, have a look at the multiple sources in lean literature (e.g., Liker 2004; Bicheno and Holweg 2008; Coletta 2012).

For this article, we will only focus on the structure of different kinds of lean events. And this will not be classical lean events but TRIZ events, which injects more innovation into the event-driven lean product development process.

5 TRIZ Events as Lean Events

TRIZ is very suitable as innovation accelerator for event-driven lean product development, because its systematic and structured approaches fit very well into the scheme of the lean events. For an advantageous integration of TRIZ events, in general, two steps have to be done:

1. Identification of the right place in the event-driven process, where some kind of TRIZ event adds innovative value
2. Structuring of each kind of TRIZ event in an appropriate way and choosing suitable tools that fit the characteristics of lean events

The starting point for the following discussion is the event-driven process shown in Fig. 1, which knows seven different kinds of lean events. For a detailed description of the schedule and tools of each kind of event, see Mascitelli 2011. But this event structure is just a template that can be transformed by each company in the way that fits best to its needs.

Regarding step 1, "identification of the right place in the process," it is necessary to identify the points in the process, where a higher amount of innovation works can increase the value of the solution most likely and with the help of TRIZ. But at the same time, the lead time should not be extended essentially. As a rough guide, one can say: the more new knowledge has to be created, the more the insertion of a TRIZ event seems reasonable.

Regarding step 2, "the structure of the TRIZ Events," one can have a look at the TRIZ tools and methods he or she already uses. Normally, these should show up a structured and maybe algorithmic nature. With this in mind, the following procedure helps to define the methodological structure of the kind of TRIZ event to be created:

- Define the goal for this kind of events.
- Collect the TRIZ tools and methods you ordinarily use to reach this goal (as a beginner, there is no other chance, than asking an experienced person for support).
- Evaluate the selected tools and methods regarding their ability to be spread over the three phases of lean events: preparation phase, actual event, and follow-up phase.
- Create the standardized tools for conduct and documentation.
- Simultaneously check the used tools, documents, methods, and approaches to fit with each other—if there are different TRIZ events designed, the variety of tools, methods, and documents should be as small as possible.

As described above, there is no paradigmatic prescription what TRIZ tools should be used. The practical experience of the authors shows that experienced TRZ users should use the tools they are most familiar with, if several tools are still on the shortlist.

In this article, mainly TRIZ tools and methods originated and inspired by Boris Zlotin, Alla Zusman, Svetlana Visnepolschi, Vladimir Proseanic, and other members of the ideation research have been chosen, because they provide a widespread range of TRIZ solutions based on tools that are very similar in structure, appearance, notation, and documentation.

5.1 Innovation Events

The purpose of the innovation events is to create problem solutions, which are very innovative. In some cases, the problems to be solved are obvious; in other cases,

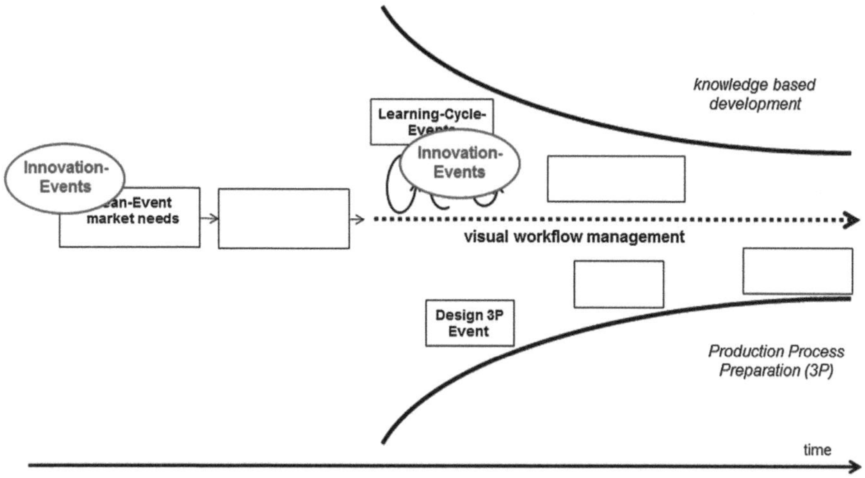

Fig. 2 Typical positions of innovation events

even the right problems have to be found first. Typically, this kind of TRIZ events fits at two places in the event-driven lean product development process (see Fig. 2).

Innovation events create innovative problem solutions and if necessary question and problems that lead to innovative solutions. Innovation does not happen accidentally, but it is the result of a systematic approach. And this result represents a gain of new knowledge for the company. The TRIZ problem-solving does not only lead to one solution, but it leads to several potential solutions—so also the solutions which are not chosen represent new acquired knowledge for the company. Because of this, innovation events can typically be placed:

- Before the market requirements lean event, to create innovative product ideas
- Alongside the learning cycle events

The market requirements event focuses on customer needs. The consideration of the voice of the customer leads to the definition of the required features of the new product. The innovation event before may be held, if there is no innovative product idea or if this idea is still fuzzy. The result of the innovation event then is the input for the market requirements event.

In the further illustration of the innovation event, the position alongside the learning cycle events is base for the description. The number of learning cycles is not specified. "These events are the answer to recognized risks based on identified knowledge-gaps. According to the principles of the knowledge-based engineering the methods used in these events are based on experiments to gather information about distribution and variation of parameters of components, materials, procedures and so on" (Thurnes and Zeihsel 2014). The innovation events have primarily the same goal: closing knowledge gaps. But in contrast to the learning cycle events whereby team is mainly closing gaps regarding the state of the art, teams are here searching for solution on higher level of innovation.

The chosen method is an adaption of the process of inventive problem solving (IPS), created by Boris Zlotin (see, e.g., Terninko et al. 1998 or Zlotin and Zusman 2004). A more detailed description of the adjustment of the IPS to fit with the characteristics of a lean event can be found in Thurnes and Zeihsel 2014—the description below explains the essential components and procedures.

Preparation Phase

The first part of the preparation phase starts with the definition of the problem, because solving this problem is the goal for this event. Typical problems may sound like "We don't know ways, how to realize the needed function" or "We have a good feature, but the realization would weaken our product" or similar. So the problem in fact is a knowledge gap, and the goal is to close it by creating appropriate new knowledge.

The Innovation Situation Questionnaire® (ISQ®) is a heuristic tool in form of a structured interview that asks several questions to gather all information needed for the team, to start the event and to raise all participants on the same level of understanding the situation (for more detailed information about the ISQ®, see Terninko et al. 1998). The checklist covers topics like a brief description of the identified knowledge gap, impacts of the problem, vision of the ideal solution, resources of the system and the supersystem, reasons for the choice of the problem, similar problem solutions already known, and analogies. This innovation checklist should be done before the event—it can be sent around the participants and when finished shared with all team members. The facilitator decides if the list is filled in completely or not. If not, he or she facilitates more rounds of filling the innovation checklist with information.

Actual Event

The duration of the actual event should be 1 day. If more time seems to be required, you should size smaller problems and conduct more than one innovation event and close multiple knowledge gaps.

To develop the innovation tasks for the team, the problem has to be defined well, using a model. This model describes specific partial problems within the overall problem. In this article, we mention using the "problem formulation" (see, e.g., Terninko et al. 1998; Zlotin and Zusman 2004) as the method to model the system and the problems within. Besides the ability to be used by a team (on a corkboard or with software support and beamer), this method provides some important advantages:

- "Documenting information about the problem and the mechanism causing it
- Developing a detailed description of the system in both its static and dynamic state
- Modeling the situation—convert the team members understanding of the situation into a visual cause-and-effect model
- Formulating an exhaustive set of "Directions for Innovation" based on the model" (Thurnes and Zeihsel 2014)

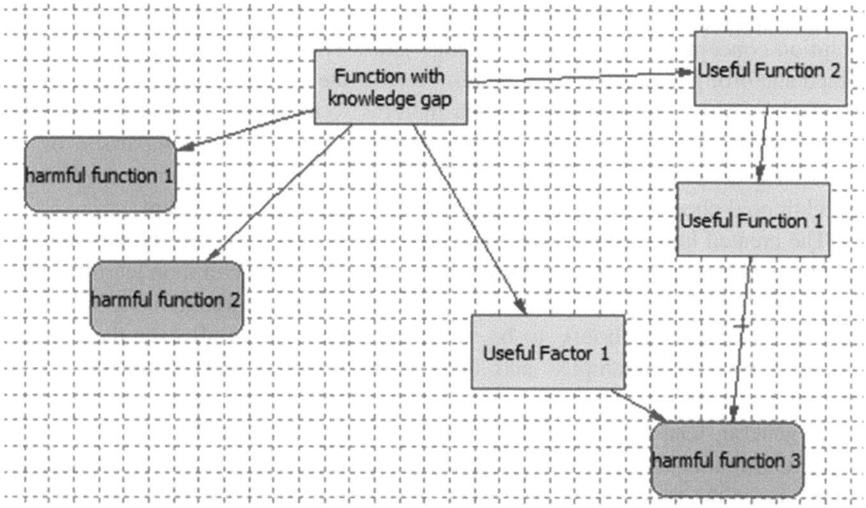

Fig. 3 Generic system diagram to model a knowledge gap (depiction created with Software Innovation Workbench®)

Figure 3 shows a system diagram as an example of the result of conducting a problem formulation. The rectangle boxes contain useful functions, conditions, or states, and the boxes with the round edges show harmful ones.

The system diagram in this case can be interpreted with an easy to perform algorithm, so "direction for innovation" in the form of smaller inventive tasks can be derived. Some examples of directions for innovation are given below:

- "Find an alternative way to obtain < Useful Factor 1 > that offers the following: does not cause < harmful function 3 > does not require < Function with knowledge gap > .
- Try to resolve the following contradiction: The useful factor < Function with knowledge gap > should be in place in order to provide or enhance < Useful Factor 1>, and should not exist in order to avoid < harmful function 2 > .
- Find a way to eliminate, reduce, or prevent < harmful function 2 > under the conditions of < Function with knowledge gap > .
- ... and many more ..." (Thurnes and Zeihsel 2014)

These directions for innovation are used as detailed tasks. TRIZ knowledge bases and tools now are sources of impact, to brainstorm for solution ideas. These knowledge bases may be heuristical hints or collections of typical technical solutions, as well as collections of physical or other effects and many others that can be found, e. g., in literature about classical TRIZ (e.g., Altshuller et al. 1999). For the purpose of the innovation events described in this article, the authors recommend using the "system of operators" (see, e.g., Terninko et al. 1998) that provides specific brainstorming hints for each direction of innovation.

After getting and evaluating solution ideas, these ideas have to be leveraged into solution concepts to close the knowledge gaps. Therefore ideas have to be combined and brought together with the resources of the system. Maybe some conflicts (in TRIZ terms: contradictions) appear thereby, so the TRIZ tools and knowledge bases can be used once more, to resolve these conflicts. A comparison of the solution concept with laws/trends and/or lines of technical evolution can be used as additional check and further improvement of the solution concept.

The created knowledge of the innovation event exists not only in the diagrams and brainstorming lists created during the event. As done in common learning cycle events, the new knowledge gained has to be stored and prepared in the so-called A3 reports or knowledge briefs, to be saved for further usage all over the learning organization (find examples and detailed information, e.g., Matthews 2011; Mascitelli 2011).

In general, lean events should show results after the actual event. But maybe some additional checks, tests, or experiments are necessary, before the new knowledge can be presented in an A3 report. In this case, an action item list has to be prepared with tasks left over for the follow-up.

Follow-Up Phase During the follow-up phase, the open tasks have to be closed. Sometimes subsequent tasks appear that way, so they have to be fulfilled or—if they show up essential knowledge gaps—be considered as topic for another innovation or learning cycle event.

5.2 Evolution Event

The purpose of the evolution event is to define promising future concepts by considering the evolution of the system—whereby the system is the product or the supersystem of the product. This approach is not just a forecast, but it respects different trends or patterns of the evolution of (mainly but not exclusively) sociotechnical systems to support the decision, in which product or service ideas may be the most promising (Fig. 4).

The innovative power of this kind of TRIZ event typically is useful to decide about the direction of upcoming systems. Regarding an event-driven lean product development process, typical positions for integration should be:

- At the very beginning
- At the end of the process

The evolution event at the beginning of the process should be organized, when the very first and raw collection of ideas for the system to be developed was created. Especially small- and medium-sized companies have no enough resources to place a bet on many horses. The evolution event helps to clearly find promising ways to go, based on an evolutionary perspective. Then the lean event market needs will use the results to exactly attune the features to the voice of the customer.

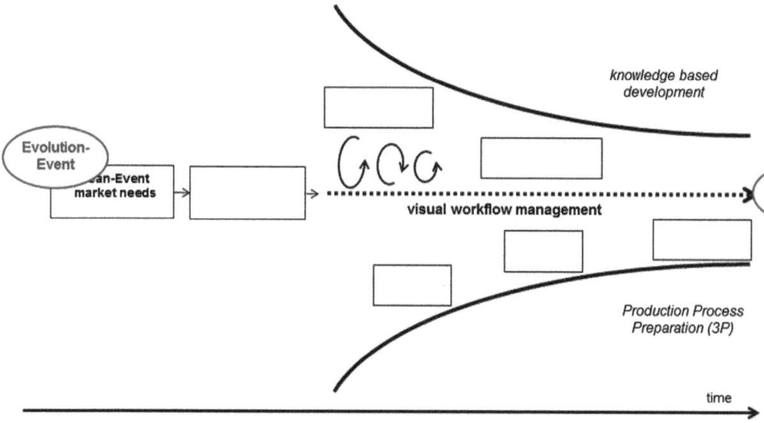

Fig. 4 Typical positions of evolution events

At the end of the process, the event confirms the decisions made and helps to define the opportunities for further development of the system.

There are several considerations about the evolution of systems in TRIZ. The description below refers to the Directed Evolution® approach (see Zlotin and Zusman 2004; Zeihsel et al. 2013). For the integration in the event-driven lean product development process, this huge approach is used in a simplified express form. And this express Directed Evolution® is portioned into pieces, meeting the requirements of lean events. The questions to be answered by the events are, e.g., "What characteristics of the product are logical in the next steps in product evolution?" or "What will be the next direction to guide the development of a successful product?" or similar.

Preparation Phase

Like the preparation for the innovation events, the preparation of the evolution event starts with a structured checklist—respectively structured interview forms. The checklist asks for the initial situation to explore the real topics of the event. An important point at this stage is also to pre-evaluate the innovative potential of the project. The evaluation encompasses different factors:

- The technical level the new product could/should reach (especially novelty and permitted or limited changes to existing systems)
- Evolutionary potential of the system (especially the system's position on S-curve and the evaluation regarding accordance to several patterns of evolution of technical systems)
- Commercial aspects (especially invention-, market-, and patent-related factors)

The checklist has to be prepared, before the event starts. The facilitator has to collect all the pieces of information. He or she arranges and documents it and is responsible for the completeness of the filled in list. Without any doubt, software support may be useful in this stage for documentation and communication.

The finished list has to be recognized by the participants of the actual event before it begins.

Actual Event

The duration of the event should be 1 day—to make this happen, the conscientious conduction of the preparation phase is essential.

The first action of the event is to formulate the ideal vision; therefore, the ideal solution is to define what is attainable with minimal changes of the existing system. This is the so-called "mini" problem.

After this definition, the main problems to be solved with future generations of the system can be identified, and some first-solution ideas can be developed.

Now, the problem and first-solution ideas are evaluated using different schemes that reflect different aspects of evolution of technical and social systems. One of these aspects is described by the TRIZ-based patterns of evolution (see, e.g., Zlotin and Zusman 2004):

- Stages of evolution—S-curves
- Evolution toward increased ideality
- Nonuniform development of system elements (contradictions)
- Evolution toward increased dynamism and controllability
- Evolution toward increased complexity followed by simplification
- Evolution with matching and mismatching elements
- Transition to micro/multi-level
- Evolution toward decreased human involvement

Each of these patterns provides hints to judge on an empirical base which of the problems and first-solution ideas found fit best to a typical and promising next generation of the system. Furthermore this approach leads to additional solution ideas and refines all the ideas. In a similar way, further trends regarding changes in lifestyle, social life, culture, markets, products, and technologies can be considered.

The facilitator then wraps all the ideas and information up and facilitates the combination of matching ideas to solution concepts together with the team.

Follow-Up Phase

Main steps in the follow-up phase are to document the solution concepts chosen as well as the ideas rejected. Maybe some subsequent tasks like experiments are needed; then their execution also is part of the follow-up. At the end, the collection of rejected ideas should be made available in the knowledge base of the company in the form that is used there typically. The elaborated solution concepts have to be given as an input to the lean event market needs or to the product planning department or the management as input for thinking about starting further development projects.

5.3 Anticipation Event

The purpose of the anticipation event is to strengthen the usual risk mitigation by adding creative components to the usual formalistic procedures like, e.g., failure mode and effects analysis (FMEA). Especially in the case of new technologies or applied new technical functions or principles, the risks are not easy to cover with usual methods that besides the systematic approach mainly refer to the team's experience. For example, the results of an FMEA depend very much on the gathered knowledge of the team members, because they brainstorm to find possible failures in single-process steps or of single components of the product. The results of this kind of brainstorming as well as the evaluation of possible failures are mainly based on the previous experiences of the team members. But the future problems may be located outside the range of actual experience—in this case, the TRIZ tool of preventive anticipatory failure determination is highly recommended. In a situation when a nagging shortcoming hinders production or troubles product performance for years, the AFD failure analysis can help to find all the critical answers regarding the underlying causes and mechanisms of the issue (AFD, see, e.g., Kaplan et al. 1999; Visnepolschi 2008). If a more formal approach should be needed all the way, a combined approach like the failure mode and effects anticipation and analysis (FMEAA) could be used (see Thurnes et al. 2012; Zeihsel et al. 2012).

The anticipation event is positioned typically in parallel to the project planning and risk mitigation event (see Fig. 5). If the anticipation event is conducted, it delivers additional risks in the form of new knowledge gaps that can be closed in learning cycle or innovation events.

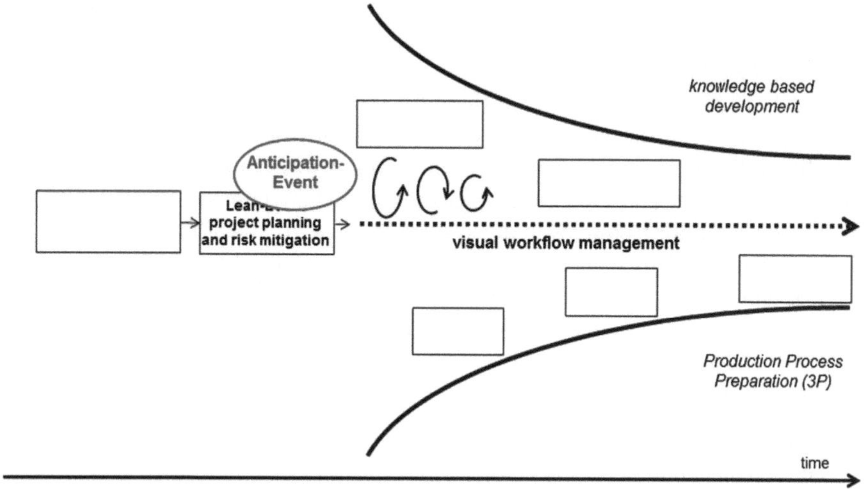

Fig. 5 Typical positions of anticipation events

To fit with the event-driven lean product development process, the original procedure has to be divided in pieces in the following way.

Preparation Phase

As for the other TRIZ events, the essential part of the preparation phase is the gathering of information. In this case, the structured interview form contains questions regarding the system of interest especially the functioning of the system, systems environment, and its history. Also undesired effects have to be enquired.

Based on this information, the facilitator develops a system diagram that shows useful and harmful functions of the system and their cause-and-effect interconnections. "Function" in this case means not only a technical function but also a condition or state. So the diagram is quite similar to the result of the problem definition in Fig. 3.

Before the actual event, the facilitator makes sure that all participants have contributed to the information gathering and show their commitment to the developed system diagram. This diagram is the main input for the event.

Actual Event

The first step of the actual event is the identification of focal points. These focal points represent zones or aspects of the system which may be responsible for the biggest harm, danger, or weakness of the system. Discussing the system diagram helps the team members understand about the focal points of the system. The facilitator can apply some rules to find focal points more easily. For example, the system diagram shows elements with relatively high number of incoming and outgoing links—that means that these elements are very important and possibly sources of big risks.

When the focal points are defined, the team starts generating failure hypotheses with some TRIZ tools. At this point, the source, effect, object, and result model (SEOR model) is very helpful to formulate harmful incidents. As important part of the AFD, this potential incident is amplified to "invent" very harmful potential incidents (see Fig. 6).

For example, a heating device (harmful source) placed nearby a plastic gear (object) may melt the gear and loses its ability to transmit torque. One of the amplified directions in this example sounds "Find ways to strengthen the heating devices to melt the pad even more." These amplified directions now are the starting points to search for resources in the system or its environment that may make this incident happen. This search can be supported using special checklists with examples of typically hazardous materials, processes, or individuals.

Failure hypotheses and failure scenarios as combination of various failure hypotheses have to be evaluated regarding the risk they represent. Typically the evaluation of these invented hypotheses and scenarios is based on:

- Grade of hazard: here the definitions of the other risk mitigation procedures may be used as common for the company.

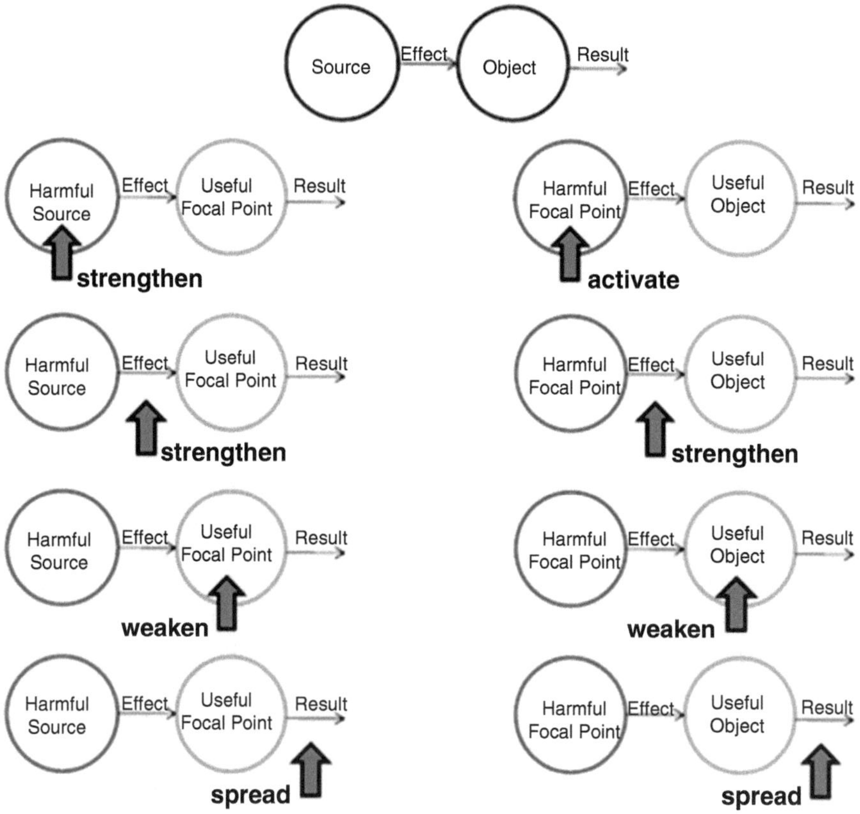

Fig. 6 SEOR configuration (see Visnepolschi 2008)

- Likelihood: in a TRIZ manner, the likelihood is very high, if the resources that may cause the theoretically possible incident are available—in general or maybe just under specific conditions.

Failure scenarios and hypotheses evaluated as very hazardous and very likely should be documented as topics for learning cycle or innovation events. The last step of the actual event is to make a decision which failure scenarios and failure hypotheses should be treated which way—this is documented in the action item list for the follow-up phase.

Follow-Up Phase

The follow-up of this event comprises the further work on the failure hypotheses and scenarios. Dependent from their nature and their evaluated impact, they can be treated in different ways:

- Pick up those hypotheses in terms of a task in the regular project plan that are evaluated as risk but without knowledge gap

- Dedicate a specific learning cycle event to those hypotheses or scenarios that are evaluated as moderate risk and knowledge gaps
- Dedicate a specific innovation event to those scenarios that are evaluated as severe risk and knowledge gap

5.4 *Intellectual Property Event*

The primary purpose of the intellectual property events (IP events) is to save the developed intellectual property from being circumvented by competitors easily. Therefore the procedure applied in the event tries to enhance the own patent—patents around the system to be developed. This leads to further ideas for stronger patents. As a result, there are a few ideas for patents besides the original idea for a patent. Usually not all these are used later, but the filing of all these patents protects the actual patent in use. Sometimes the Patent Deconstruction® approach (see Proseanic and Visnepolschi 2011) leads to the decision that one of the new patent ideas will be realized instead of the one initially planned. The Patent Deconstruction® approach is also applicable in challenging strength of a competitor's patent that hinders development of the company product or process. In this case, the task is to modify the technology in question so that it becomes simpler, more reliable, and easily implementable. The procedure typically results in finding well-known and infringement-free alternatives to the problematic technology and in truly original ideas that are often patentable as well. The TRIZ approach for enhancing patents that way is not a legal approach—instead of thinking about enhancing the strength of the formulation of the claims, it is suggested to invent claims for stronger technical concepts (Fig. 7).

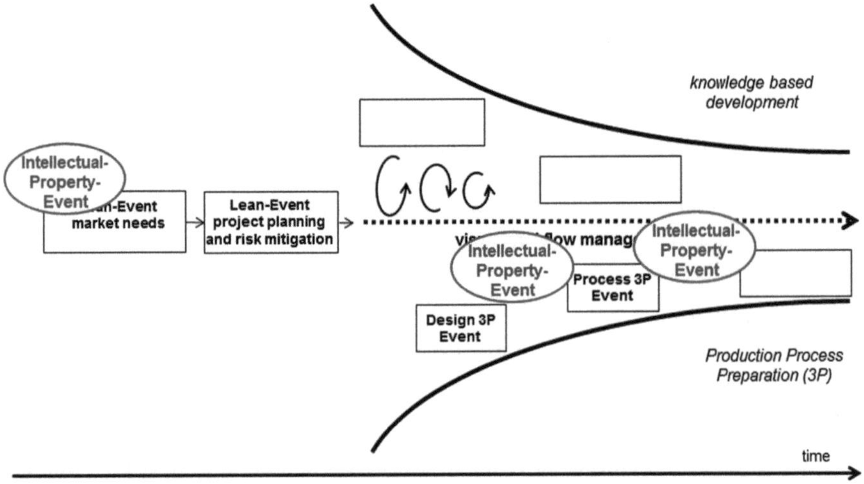

Fig. 7 Typical positions of intellectual property events

IP events may take place at any position in the event-driven lean product development process. Especially at the beginning of the process as well as around events that create "new reality" in product or process, they may be very useful. So each company using this approach has to decide if there are some standard IP events (e.g., at the beginning and the end of the development process) or if there are exceptional IP events requested by the project leader or a specified committee or if the company needs a combination of both.

To keep the plenty of used methods and models small, also in this case, the modeling notation of problem formulation is used (similar as mentioned above in other event descriptions, see Proseanic and Visnepolschi 2011).

Preparation Phase

To prepare an IP event, the structured interview form to gather initial information asks for some details of the invention patented (or to be patented). Besides the name, especially an entire description of the claims should be available. Furthermore it will be good to know about competing patents in the same field.

Actual Event

To prepare further ideas for strong patents, first the initial patent (or patent idea) is modeled on a corkboard or with an appropriate software tool. Figure 8 shows an example for such a model. Essential parts of this model are taken out of the description of the independent claim of the patent (or patent idea).

The team now works with this model, following several standardized suggestions that are based on:

- Generalizing elements of the modeled claim, to create a wider range for a new patent
- Trimming away parts of the invention, to create a new system with less elements but at least the same useful functioning as the original system

To find the right ideas, standardized TRIZ tools may be used—as described earlier, we remain with the system of operators, because it is very effective and also used in the other events. Figure 9 shows an interim result of working on the patent model shown in Fig. 8.

As a result, ideas for new patents are documented and formulated as independent claims. Maybe in some cases, the team doesn't find a way to realize the new model easily—then a shortened version of the innovation event can be embedded, to close the identified knowledge gap inventively.

Follow-Up Phase

The follow-up in this case contains meeting the decision, in which ideas from the list actually should be filed. Then the further process is consigned to the respective department or person. In some cases, the responsible persons may decide to publish some of the ideas, to realize freedom to operate.

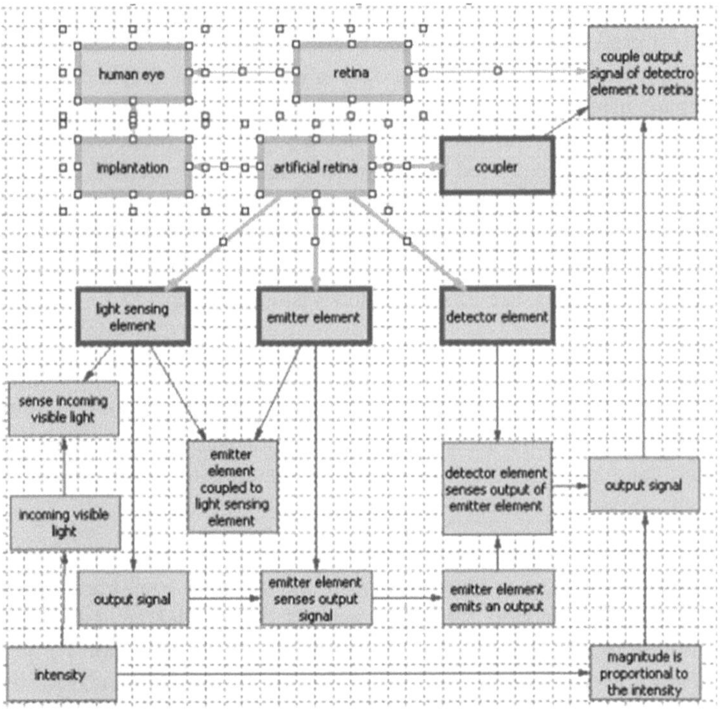

Fig. 8 Model of an independent claim of a patent (depiction created with software "invention enhancement," by kind permission of Vladimir Proseanic)

6 Conclusions

Based on the very pragmatic and efficient event-driven lean product development process, this article shows a useful way to add innovative power at distinct stages. Advantages using the TRIZ events as described are:

- Adding very systematic and thereby efficient elements to create more innovation and higher quality
- Having a need-oriented event set, to gain efficiency
- Getting sophisticated TRIZ approaches in a structured action procedure that is compatible with the characteristics of lean events
- The concentrated selection of just a few TRIZ tools does not confuse the users

So far the concept described in this article is partly practical proven. The TRIZ tools are well known and state of the art—their adaption to the lean event structures and requirements is still under development. The authors already gained experience in performing several TRIZ events and refine the concept actually. The usage of the TRIZ events all in all leads to the development of more new knowledge per time. And this knowledge has a higher quality regarding its level of innovation. Therefore

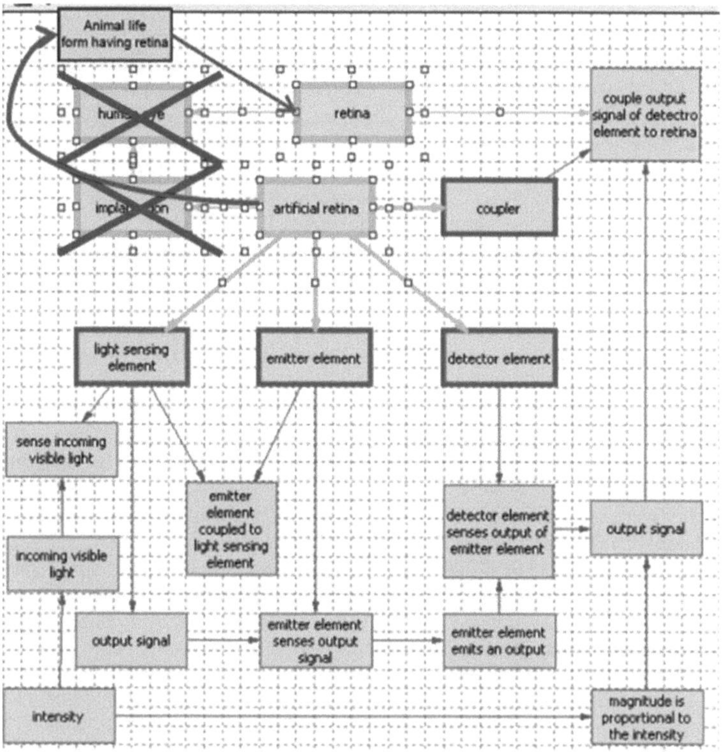

Fig. 9 Model of an independent claim of a patent with modifications to develop stronger patents (depiction created with software invention enhancement, by permission of V. Proseanic)

the practical event structuring of TRIZ applications is a promising approach for the further improvement of efficient product development processes.

References

Altshuller, G. S. (1984 [1987]). *Creativity as an exact science*. New York: Gordon and Breach Science.

Altshuller, G. S. (2006). *And suddenly the inventor appeared*. Technical Innovation Center. Seventh Printing, Translated by Lev Shulyak. ISBN 0-9640740-2-8.

Altshuller, G. S., Zlotin, B., Zusman, A., & Philatov, V. (1999). *Tools of classical TRIZ*. Southfield, MI: Ideation International.

Bicheno, J., & Holweg, M. (2008). *The lean toolbox: The essential guide to lean transformation* (4th ed.). Buckingham: PICSIE Books. ISBN 978-0954124458.

Coletta, A. R. (2012). *The lean 3P advantage: A practitioner's guide to the production preparation process*. Boca Raton, FL: CRC Press.

Hiltmann, K., Souchkov, V., Thurnes, C. M., Adunka, R., Koltze, K., Livitov, P., et al. (2014, September 4–6). *Standard VDI 4521: Solving inventive problems with TRIZ* (S. 247–254). In: The International TRIZ Association—MATRIZ (Hrsg). The 10th International Conference TRIZfest-2014, Theories and applications, Conference proceedings, Prague.

Kaplan, S., Visnepolschi, S., Zlotin, B., & Zusman, A. (1999). *New tools for failure and risk analysis: An introduction to Anticipatory Failure Determination (AFD) and the theory of scenario structuring*. Southfield, MI: Ideation International. ISBN 1-928747-05-1.

Liker, J. (2004). *The Toyota way: Fourteen management principles from the World's greatest manufacturer*. New York: McGraw-Hill Professional. ISBN 978-0071392310.

Mascitelli, R. (2006). *The lean product development guidebook: Everything your design team needs to improve efficiency and slash time-to-market*. Technology Perspectives. ASIN B00DT67ZZA.

Mascitelli, R. (2011). *Mastering lean product development—a practical, event-driven process for maximizing speed, profits, and quality*. Quality Books. ISBN 13-978-0-9662697-4-1.

Matthews, D. (2011). *The A3-Workbook*. Boca Raton, FL: CRC Press.

Proseanic, V., & Visnepolschi, S. (2011). *Improving patents and designing around blocking patents using PATENT DECONSTRUCTION® method*. TRIZCON, Detroit. Proceedings available at http://www.aitriz.org/documents/TRIZCON/Proceedings/2011-10_Applying-Patent-Deconstruction.pdf

Terninko, J., Zusman, A., & Zlotin, B. (1998). *Systematic innovation: An introduction to TRIZ*. Boca Raton, FL: CRC St. Lucie Press. ISBN 1-57444-111-6.

Thurnes, C. M., & Zeihsel, F. (2014, June 8–11). *Innovation-events enriching efficient lean product development processes*. In: ISPIM–International Society for Professional Innovation Management (Ed.), Proceedings of the XXV ISPIM Conference 2014, Dublin, Ireland. Proceedings available online at www.ispim.org

Thurnes, C. M., Zeihsel, F., Visnepolschi, S., & Hallfell, F. (2012, June 24–26). *Using TRIZ to invent failures—concept and application to go beyond traditional FMEA* (pp. 263–290). In: Proceedings of the TRIZ Future Conference 2012, Lisbon, Portugal, 17 pages.

Visnepolschi, S. (2008). *How to deal with failures (The smart way)*. Farmington Hills, MI: Ideation International. 204 pages.

Ward, A. (2007). *Lean product and process development*. Cambridge, MA: Lean Enterprise Institute. ISBN 978-1934109137.

Zeihsel, F., Thurnes, C. M., Zlotin, B., & Zusman, A. (2013, June 16–19). *Guided innovation by using directed evolution® for next product generation* (15 pages). In: ISPIM—International Society for Professional Innovation Management (Hrsg). Proceedings of The XXIV ISPIM Conference—innovating in global markets, Helsinki, Finland.

Zeihsel, F., Thurnes, C. M., & Visnepolschi, S. (2012, June 17–20). *New risk assessment for innovation management* (14 pages). In: ISPIM–International Society for Professional Innovation Management (Hrsg). Proceedings of The XXIII ISPIM Conference—action for innovation: Innovating from experience, Barcelona, Spain.

Zlotin, B., & Zusman, A. (2004). *Directed evolution®: Philosophy, theory and practice*. Southfield, MI: Ideation International. 105 pages.

Advanced Function Approach in Modern TRIZ

Oleg Feygenson and Naum Feygenson

Abstract In this chapter, the authors describe Advanced Function Approach (AFA) which was introduced as an analytical tool of Modern TRIZ in 2010. At that time, it was shown how utilizing the spatiotemporal parameters can further enhance such a powerful analytical tool as Function Analysis for Engineering Systems.

Since then, AFA has proved its practical efficiency—dozens of systems have been improved or developed using the results of AFA. Moreover, methodological recommendations for applying AFA have been developed and verified in the following areas:

- Revealing and describing the synergetic effect of combining two Engineering Systems;
- Specifics of Function Analysis for Engineering Systems at the exploitation stage;
- Novel approach to categorizing functions that Engineering Systems perform;

This chapter also reflects milestones in developing Function Analysis and incorporating it in Modern TRIZ.

Keywords Function • Function Approach • Spatio-temporal parameters

1 Introduction

Lawrence D. Miles, the universally recognized founder of the function approach, developed Value engineering (VE)—a systematic method to improve the "value" of goods or products and services by examining their functions. By definition, value is the ratio of function to cost. VE follows a structured thought process that is based

O. Feygenson (✉)
MP Solving, Brussels, Belgium
e-mail: oleg.feygenson@gmail.com

N. Feygenson
TRIZ-InCon Company LLC, Saint Petersburg, Russian Federation
e-mail: feyg11@gmail.com

L. Chechurin (ed.), *Research and Practice on the Theory of Inventive Problem Solving (TRIZ)*, DOI 10.1007/978-3-319-31782-3_12

exclusively on "function," i.e., what something does, not what it is (Kaufman 2009).

The history of developing function approach is described in detail in (Litvin et al. 2010). Nowadays, about 20 ways to define and utilize the concept of "function" in various fields of science and engineering exist. Those concepts are described and discussed in (Houkes and Vermaas 2010). Results of extensive research on philosophical and methodological aspects of applying the function approach can be found in (Vermaas 2013; Cummins 1975; Vermaas 2010). Some methodology behind the application of the function approach for enhancing cause-and-effect analysis is described in (Chittaro and Kumar 1998; Weber and de Preester 2005).

Here, in this chapter we will focus on the practically oriented function approach and the function analysis that are used in Modern TRIZ for analyzing engineering systems.

Function analysis, as defined in Modern TRIZ, is an analytical tool that identifies functions, their characteristics, and the cost of System and Supersystem components (Gerasimov and Litvin 1991). This type of analysis is needed to identify the system's disadvantages such as harmful functions, insufficiently or excessively performed useful functions, excessive cost of components and, as was proposed recently, the wrong place and time of performing functions, and the absence of the required functions (Litvin et al. 2010). Another goal of Function Analysis is to prepare a model of the system to be used in the subsequent stages of the analysis: Flow Analysis, Cause-Effect Chain Analysis (CECA) or Root-Cause Analysis (RCA), Trimming, Feature Transfer, and Super Effect Analysis.

2 Essence of Advanced Function Approach

The function analysis for engineering systems, as developed by specialists of the Leningrad (St. Petersburg) TRIZ school, Vladimir Gerasimov and Simon Litvin, is characterized with the following attributes:

- A concrete, practical definition of function was introduced: "An action performed by one Material Object to change or maintain at least one Parameter of another Material Object."
- A triad for the description of a function was suggested: "function carrier—action—object of the function."
- Rules and algorithm for accurately formulating functions were developed.
- A parametrical evaluation of the level of function performance was suggested.
- Also, the concepts of harmful and neutral functions were introduced.

Graphically, a function can be represented as shown in Fig. 1.

The specific steps and detailed procedure of the Function Analysis are described in the classical work of Vladimir Gerasimov and Simon Litvin published in Russian (Gerasimov and Litvin 1991). In English, some highlights can be found in (Litvin

Fig. 1 A graphical representation of function

et al. 2010). Here, we would like to describe briefly the procedure of the Function Analysis as it is adapted in TRIZ and then demonstrate how AFA can enhance this analytical tool.

When we are analyzing an Engineering System, the first thing we have to do is to split this system into components. This part of Function Analysis is called a Component Analysis. As a result of this procedure, we will have a list of components that make up the system. The list should also include several components of the supersystem with which our system interacts. Those supersystem components can be used as resources in the future when we improve the system (e.g., they can be used as new potential function carriers if we decide to trim components of the system).

Once the list of components has been composed, we need to identify interactions between those components. If two components physically interact (touch each other), then there is a chance that they each perform some function. There can be cases when Component A performs a function for Component B, while Component B performs another function for Component A. For example, a cup (Component A) holds coffee (Component B), and at the same time, coffee (Component B) heats the cup (Component A).

When all components and all interactions between those components are identified, we have to formulate functions that may take place between components which interact.

According to Gerasimov and Litvin (1991), "there is a certain sequence to defining a useful function of a function carrier:

1. Suggest an initial formulation of a function that seems correct.
2. Ascertain whether the function carrier could perform the proposed function itself (the criterion for this is the presence of at least one element in the carrier that participates in the execution of the function).
3. Formulate a more precise definition of the function by asking the following questions: "what is the purpose of performing the function?" (if the element mentioned in #2 above is evident); "how exactly is the function performed?" (if the element in #2 is not evident).

If the initial function is imprecise, procedures 2–3 should be repeated until a precise definition is found. The criterion for finding a meaningful, precise definition is that at least one element of the object being analyzed takes part in performing the function."

AFA proposes adding the following steps to those above (Litvin et al. 2010):

4. "Indicate the place the function is performed;

The indication of place should be precise and specific, because the same function could have very different levels of performance in different places.

5. Indicate the time the function is performed."

These two additional steps seem simple, but they really help to understand the system and identify its disadvantages; let us consider a case study presented in (Litvin et al. 2010):

Cleansing teeth with a toothbrush.

This example with a toothbrush was used by Simon Litvin and Vladimir Gerasimov when they were developing the TRIZ-based Function Analysis. Nowadays, this example is widely used in teaching Function Analysis at TRIZ seminars. The Function Model of toothbrush was also used by Sergei Ikovenko when he prepared his Master TRIZ Thesis "TRIZ Application for IP Strategies Development" (Ikovenko 2006).

Obviously, the main function of the engineering system under analysis is "to remove plaque (from the teeth)." A fragment of the function model at the top hierarchical level is depicted in Fig. 2.

According to Litvin et al. (2010): "The model states that the basic function of the toothbrush "to remove plaque" has an insufficient level of performance, assuming that such level was estimated by the parameter "amount of dirt." This is all of the information we identified about the interaction of two components: the toothbrush and the plaque.

However, the important fact that plaque between teeth is removed insufficiently is not stated by the function model. Actually, according to dentists' recommendations it is necessary to use floss to complete a cleansing procedure.

Moreover, the most dangerous location of plaque is the area under the gum, which is where parodontosis starts. The very important fact that the toothbrush does not remove plaque from the area under the gum is not indicated.

If we consider ES toothbrush as a set of functions which are characterized by the location and time of performance, then it becomes clear that the toothbrush removes plaque from different surfaces of the teeth and, which is especially important, at different moments in time." The fragment of Function Model built with AFA is shown in Table 1.

By using the Advanced Function Approach, it is easy to show that, in reality, the interaction between two components "toothbrush" and "plaque" is described by at

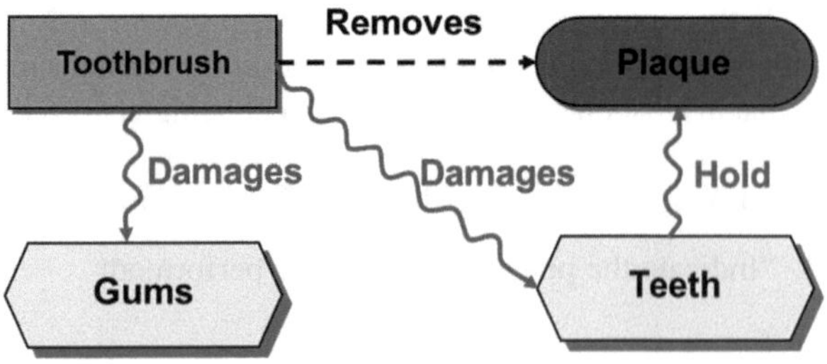

Fig. 2 A fragment of the function model at the top of the hierarchy

Table 1 A fragment of the function model built with AFA

Time	Function			Type/rank and level of performance	Location	Comments
	Function carrier	Action	Object of function			
t0	Teeth	Hold	Plaque	Harmful	Whole surface of teeth: external, chewing, palatal, between teeth, under gum.	
t10	Toothbrush	Removes	Plaque	Basic, normal	External side of teeth	
t20	Toothbrush	Removes	Plaque	Basic, normal	Chewing side of teeth	
t30	Toothbrush	Removes	Plaque	Basic, insufficient	Palatal side of teeth	
t40	Toothbrush	Removes	Plaque	Basic, insufficient	Side surface of teeth (between teeth)	
t10–t40	Toothbrush	Damages	Teeth	Harmful	Whole surface of teeth	
t10–t40	Toothbrush	Damages	Gums	Harmful	Near teeth	Different sides of gums could be considered as well: external, between teeth, etc.
...	...	...	...	...	...	

least four derivatives of a function, which are performed in different places and at different moments of time. It is also important that the levels of performance of these derivatives are different:

1. Toothbrush—"removes plaque" (time 10—the beginning of tooth brushing; external side of teeth); adequate level of performance;
2. Toothbrush—"removes plaque" (time 20—in 45 s after beginning of tooth brushing; chewing side of teeth); adequate level of performance;
3. Toothbrush—"removes plaque" (time 30—in 90 s after beginning of tooth brushing; palatine side of teeth); insufficient level of performance;
4. Toothbrush—"removes plaque" (time 40—in 135 s after beginning of tooth brushing; between teeth); insufficient level of performance.

It is also identified that the harmful function "teeth hold plaque under gum" exists, but there is no useful function "toothbrush removes plaque under gum."

Let us compare the results obtained by using the classical TRIZ-based Function Approach and the results from applying Advanced Function Approach.

- While using the classical TRIZ approach a functional disadvantage was detected and documented: the level of function implementation "toothbrush—"remove dirt" is insufficient (for example, by the parameter "amount of dirt")
- With the use of the Advanced Function Approach it was clearly demonstrated that:
  - The toothbrush adequately removes dirt from the external and grinding surface of teeth.
  - The toothbrush insufficiently removes dirt from the inner and flank surface of teeth.
  - The toothbrush does not remove dirt from under the gum. In other words, a very important function, which is required of the ES, does not exist.

Such information is essential in an analysis of a real ES, since it helps to better understand systems functionally and identify the drawbacks related to the discrepancy between the factual place of function implementation and the required one.

The function model built using AFA shows that the same function carrier "toothbrush" implements both the useful function "to remove dirt" and the harmful function "to damage teeth" in one place—"on the tooth surface (external/palatal/flank)"—and at the same time. This situation can be modeled as a Physical Contradiction: "The toothbrush has to contact the tooth surface to remove the dirt and the toothbrush should not contact the tooth surface in order not to damage it." Other weaknesses of the existing toothbrush have been identified and described as contradictions in (Litvin et al. 2010).

Based on practical experience, AFA does not take much more time and effort than the classical Function Approach to analyze an Engineering System, but AFA does allow us to find the real nonobvious disadvantages of a system under analysis. Moreover, application of AFA significantly accelerates the subsequent analytical and problem-solving procedures.

3 Synergetic Effect Achieved When Two Identical Systems are Joined. Practical Example of Applying Advanced Function Approach

Let us discuss how the Advanced Function Approach can be used to reveal and describe the synergetic effect of combining two engineering systems. The simplest scenario is when two identical systems are joined together. When building a function model of such a bi-system, it is important to know the specifications for "when" and "where" each system in the bi-system performs its function. This will give us a clear understanding of what the best regime for the conjoined operation is.

As explained in (The Strategy Reader 2004), the Synergetic Effect is an effect happening between two or more material objects, factors, or entities that produce a joint effect which is more significant than the simple sum of their own single effects.

A good illustrative description of the synergetic effect is presented in (Macnamara 2012). When we look at parts, we cannot predict the whole system and its characteristics. For example, the wetness of water cannot be predicted if we only consider the molecules or atoms of Hydrogen and Oxygen separately. So, in order to understand a system we have to understand interactions between the components of this system as well as changes in their parameters.

Certainly, people try to obtain the highest functionality possible from the systems they utilize. Modern TRIZ offers a number of tools that can help to dramatically improve the functionality of Engineering Systems by combining/merging several systems, e.g., Feature Transfer, Hybridization, Trend of Transition to the Supersystem, etc.

However, these TRIZ tools do not allow us to clearly identify whether a synergetic effect is achieved by combining two or several systems. The Advanced Function Approach can be considered a tool for revealing this synergetic effect.

Let us consider in detail how the synergetic effect can be achieved in the simplest case of combining two identical engineering systems. First of all, we need to consider all possible conditions in which the actions of the systems we are combining perform. Table 2 summarizes these conditions.

Let us now consider how the expected results depend on the conditions of actions performed.

No. 1: the results are additive, but they can be achieved in a shorter period of time.

No. 2: does not lead to reducing the time needed to achieve the results; the results are simply summarized.

No. 3: the result doubles, but the time for achieving this doubled result remains the same.

No. 4: the time needed to achieve the result is reduced, but the value of this result strongly depends on the exact interaction of the two systems in the zone of co-action. That is, the systems could even impede each other.

So, in order to make an informative conclusion about the efficiency that can be achieved when combining two systems, the first step is to understand the spatio-temporal parameters of their functions.

In order to illustrate the above approach, let us look through a case study. Imagine there are two identical excavators that are digging a ditch together. It is assumed that the ditch has a simple shape: its cross section is rectangular and there are no turns along the ditch. The width of the ditch is equal to the width of one excavator bucket.

Table 2 Conditions of performing actions when combining two systems. Possible combinations

		Temporal description of performing actions	
		Simultaneously	Sequentially
Spatial description of performing actions	In different places	1	2
	At the same place	4	3

- *No. 4:* if two buckets are directed into the same place simultaneously, then the result achieved is rather harmful.
- *Nos. 2 and 3* do not really make sense in this specific example.
- *No. 1* is the most interesting one. If the functions are performed in completely different places (i.e., the excavators dig from opposite ends of the ditch), the achieved results will simply be summarized. However, if both excavators dig simultaneously and the place where the functions are performed is properly organized, then a synergetic effect can be obtained. For example, the first excavator removes the ground from the upper half of the ditch and the second one removes it from the lower half; see Fig. 3. In this case, we will dig the ditch more than two times faster since the excavators do not need to adjust the positions of their buckets along the vertical extend. So, each of them works more efficiently.

But the opposite effect can also be achieved—when such separation of place (where the functions are performed) leads to increasing the total time of the operation. To work properly, each excavator should operate at its respective depth, and one excavator must remove ground from the ditch at a distance sufficient so as not to hinder the second. However, if the ditch is not long enough, there is no positive effect at all from combining the two excavators.

So, in order to reveal a synergetic effect and describe it in terms of functions, it is necessary to use the spatiotemporal parameters of actions. When two identical systems are combined into one system as discussed above, there was no way to explain the expected results without clarifying the two parameters "time of performing a function" and "place of performing a function." So, we have illustrated another important practical application of the Advanced Function Approach.

Fig. 3 Excavators joined into a bi-system

4 Exploitation Stage of a System's Life Cycle, Its Uniqueness, and Use of Function Analysis in This Stage

The exploitation stage for any system is the point where the system performs all of the useful functions for which it was designed. It is vitally important that the system performs these functions well and just how well it does this characterizes the overall performance of the system.

In the Innovative Consulting practice, very often we deal with systems that are in the exploitation stage. Our customers ask us to improve performance or reduce the cost of existing systems. In these cases, innovative ideas for system improvement may be found at the previous or subsequent stages, but we always start with the Function Analysis of the system at the exploitation stage.

Elegant examples of where opportunities for system improvement were revealed after a simple analysis at the exploitation stage are presented in (Ackoff 1978). One of those examples illustrates how the direction for further development of refrigerators was identified within a consulting session. During this session, it was understood that most of the times people open a refrigerator's door in order to get some cold drink or ice cubes. Is it possible to get ice or cold drink without opening a door of refrigerator? That is how the concept of ice and water dispenser built into the door was developed.

The examples from (Ackoff 1978) demonstrate that in order to reveal opportunities for system improvement we do not necessarily need to go deep into the technical details. A simple analysis at the highest hierarchical level, the Supersystem level, could be extremely effective.

Also, when we talk about system exploitation we really mean a set of actions, one after another. This is a process. Moreover, if we take a look at the list of components involved in this process, we will note that these are components of the Supersystem, not the system itself.

For example, if we take the manual for a dishwashing machine, we will read a description of the process, e.g., as follows:

> 1. After loading the dishwasher and adding detergent, select the desired cycle and options by pressing the buttons. The indicator lights for the selected cycle and options will light up. To cancel an option, press the button again.
>
> 2. To start a cycle, close the door until it latches, then press the START/Cancel button once. After a pause, the filling will begin. The display countdown will flash until START/Cancel is pressed.
>
> Etc.

So, exploitation is a process and if we need to analyze the system at the exploitation stage of its life cycle, then we need to apply Function Analysis for Technological Processes which is more complicated than the Function Analysis for Products. It utilizes different rules and ranking of functions.

This is a typical TRIZ situation—we have the engineering contradiction: IF we use Function Analysis for Products to analyze the system at the exploitation stage, THEN it is easy to implement, BUT the function model is not informative enough.

And vice versa: IF we use Function Analysis for Processes to analyze the system at the exploitation stage, THEN the function model describes the system precisely and highlights all its functional disadvantages, BUT it is difficult to implement since it requires knowledge of more advanced rules and ranking systems to formulate functions.

Fortunately, TRIZ was developed to resolve contradictions. One possible resolution for the contradiction mentioned above is to apply the Advanced Function Approach in order to analyze the system at the exploitation stage.

As was described in the previous section, AFA is based on the assumption that the spatiotemporal parameters "time of performing a function" and "place of performing a function" are the most universal when characterizing any action. In fact, any action takes place within a certain period of time and in a certain space. Utilization of the spatiotemporal parameters can enhance such a powerful analytical tool as the Function Analysis for Engineering Systems.

According to recommendations of AFA:

1. Use spatiotemporal parameters when describing an Engineering System functionally: "time of performing a function" and "place/allocation of performing a function."
2. Combine features of both types of Function Analysis:
   - Function Analysis for Processes gives us a clear understanding of sequence and duration of functions.
   - Function Analysis for Products gives us a structural picture of the analyzed engineering system—collocations and interaction of components of the system and its Supersystem.
3. Introduce a new type of function disadvantages to the Function Analysis technique. That is, disadvantages which are related to the mismatch of real time and place of performing the function with the required ones.

A practical example where AFA is used to analyze the engineering system at the exploitation stage is shown in (Feygenson 2012).

Now let us demonstrate that the Function Analysis of the system at the exploitation stage is an essential part of any project.

Because any Engineering System is designed to perform its functions and these functions are performed at the exploitation stage, it is very important to analyze this stage whether or not the scope of the project includes it. A list of typical stages may include Design, Manufacturing, Transportation, Storage, Exploitation, Maintenance, and Utilization.

Imagine that we are analyzing the system at the manufacturing or even design stage of its life cycle. It is extremely important to forecast what can happen with the system when the final consumer starts using it. It is also important to forecast what will happen with the Supersystem in which the system will exist. One of the best known examples of when a wrong forecast was made is when Bill Gates said just a few years ago that 64 kilobytes of memory is enough for anyone. How much memory are we using now in our computers? This is because the Supersystem developed far more than had been presumed: software, visual applications, etc.

The example above clearly demonstrates how important the development of the Supersystem is. When we design and develop the system we should understand that by the time the system reaches the final consumer, the Supersystem can dramatically change.

Now, let's imagine we are analyzing a system which is at the stage after the exploitation stage, i.e., maintenance or utilization. It is very important to know the system's history: how it was used, with which components it was interacting, what were the conditions of the Sypersystem, etc. All this information can help us identify the right strategy for maintenance or utilization. The Function Analysis of the system at the previous stage, the exploitation stage in this case, will provide all required information including opportunities for function redistribution and list of potential resources for the stages that follow the exploitation stage of the system's life cycle.

The main point here is as follows: the Function Analysis performed for a system at the exploitation stage can significantly enhance the efficiency of the analytical stage of the project even if the specific scope of the project does not include this stage. Application of the Advanced Function Approach will make the results of the analysis more focused and informative.

Briefly, the recommendations for Function Analysis of a system at the exploitation stage could be summarized as follows:

- The Function Analysis should be performed at the highest hierarchical level, at the level of the Supersystem. When the system is at the exploitation stage, it is very important to take into consideration its interactions with the Supersystem's components, including the object of the main function—Target.
- Advanced Function Approach can significantly enhance the efficiency of the Function Analysis. In fact, if we apply the spatiotemporal parameters "time of performing a function" and "place of performing a function," we combine features of both Function Analysis for products and Analysis for processes.
- It is also recommended to perform a quick Function Analysis of the system at the exploitation stage, even if we consider any other previous or subsequent stage as a project scope:
  - At a previous stage (manufacturing, transporting, etc.), such analysis will allow us to predict and take into account all further important interactions with the Supersystem.
  - At a subsequent stage (storage, maintenance, etc.), such analysis will provide a list of actual resources that are available in the Supersystem.

Recommendations on how to perform the Function Analysis of a system at the exploitation stage are presented in more detail in (Feygenson 2012).

5 Approach to Categorizing Functions that Engineering Systems Perform

As has already been mentioned, Function Analysis is one of the most developed and formalized analytical tools in Modern TRIZ. That is why it does not require excessive time and intellectual effort from an experienced researcher to obtain a thorough understanding of how the system under analysis works and what its function disadvantages are. However, a complete list of recommendations for selecting the most appropriate function(s) is still lacking. To date, there have been several attempts to address this issue (Litvin et al. 2010; Gerasimov and Litvin 1991; Hirtz et al. 2002; Kynin and Priven 2013), but most of them are being used only by their developers. Usually, these attempts contain recommendations that make the Function Analysis procedure more complicated but do not really enhance its output.

The recent fundamental research published by Yury Fedosov (2013) recommends "the most simple and obvious way of formulizing the procedure for formulating functions" is to choose the appropriate function for the current situation from "the list of elementary functions." This empirical study (Fedosov 2013) is based on the representative statistical data taken from TRIZ consulting projects performed successfully in different engineering and scientific areas.

Another approach for formulating the functions of the system's components is based on the application of detailed algorithms. The first algorithm for formulating functions was proposed by Gerasimov and Litvin (1991). Recently, Kynin and Priven (2013) developed an algorithm for selecting the elementary functions from the matrix $4 \times 4 \times 4$, which simplifies and standardizes the procedure, thereby reducing the risk of erroneous formulations.

Hybridization of these two known approaches allows us to propose a novel approach for performing Function Analysis that is simple enough and not time-consuming, but still sufficient for describing complex engineering systems and revealing their function disadvantages.

A hypothetico-deductive method of creating scientific theories is applied. This is when a whole working theory is built upon assumptions. Basically, this method organizes knowledge that existed before and extends the range of applications where that knowledge may be applied.

The novel approach described here is based on the two following statements:

- According to recommendations of AFA, the spatiotemporal parameters should be used when describing an Engineering System functionally: "time of performing a function" and "place/allocation of performing a function" in each specific situation. This leads to increasing the accuracy of the system description. Such detailed description allows us to identify function disadvantages in the system that are difficult to observe with the classical function approach.
- The original supposition is that the vast majority of functions that engineering systems perform can be covered by four base types of functions described in

AFA format. The appropriateness of this supposition is supported by its testing with concrete empirical data.

Thus, we can formulate *the main assumption*: When the AFA format of function description is utilized, the vast majority of actions (the verb-related part of a function formulation) can be described with the following verbs:

- ***To move*** (Note: when AFA format is applied the specification of direction, speed, distance is taken into account);
- ***To hold/to stop*** (Note: the final choice depends upon the initial stage of the function recipient. For example, "to hold" can be applied if it was a stationary stage; "to stop" is more appropriate if the object was moving);
- ***To combine/to join;***
- ***To separate/to divide.***

Actions covered by the verbs above are general enough; they can be used for describing what any material objects (including substances and fields) do. This approach is illustrated in Fig. 4.

At first glance, the proposed approach may look oversimplified and insufficient for describing complex engineering systems. However, in our opinion it is sufficient because special attention is given to the parametric characterization of each function including its object, subject, and action.

The first version of the "Elementary Functions" Handbook proposed in (Fedosov 2013) has been used to validate the proposed approach. The Handbook is based on a statistical approach: 32 function models of actual complex engineering systems from different industries were analyzed. In total, 256 components performing 2132 functions were considered.

In fact, we substituted each function in the Handbook with one of the base/underlying functions proposed above and parameters of performing this function.

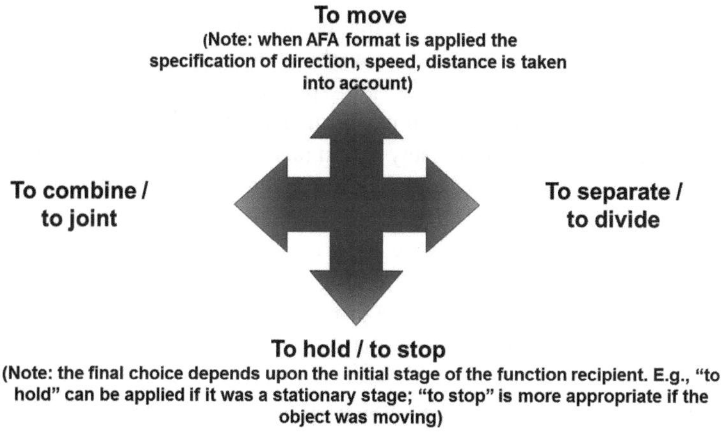

Fig. 4 Four generalized types of functions

More details of this research can be found in the recent publication (Feygenson et al. 2014).

Not only is the novel approach described here ready for application in actual projects, but it can also simplify introducing and explaining the function analysis to people who are not familiar with this analytical tool.

Basically, we can start with a general undefined description of the system we are analyzing. This will give us an initial understanding of the problem as seen by the client (a person or a project team) that owns the problem. Then, we need to create a function model by reformulating what the client knows using the set of base/underlying functions. Of course, all of the parameters of each function, including parameters of its object, subject, and action, should be taken into account.

Since the number of base/underlying functions (actions) is limited by four types and seven verbs, creating the function model does not take much time. Actually, creating a function model can be performed together with the client in real-time facilitation mode. The results of such facilitation can be vitally important for a complete final understanding of how the system works and what its function disadvantages are.

In general, the novel approach to categorizing functions is characterized by the following features:

- Simplified procedure and reduced time of the function analysis without compromising quality of its results;
- Ease of introducing/explaining the function analysis to people who are not familiar with this most important analytical tool in Modern TRIZ.

6 Summary

Based on an evolutionary analysis of the effectiveness of applying parameters for formulating functions, it was assumed that not only the object and the recipient of the function but also the action (the verb) should be characterized by some parameters. It was suggested that the spatiotemporal parameters "time of performing a function" and "place of performing a function" are the most universal to characterize any action. In fact, any action takes place within a certain period of time and in a certain space. This approach was given the name "Advanced Function Approach."

Several applications of AFA have been developed since it was introduced in 2010. Those applications lead to increasing efficiency of analyzing Engineering Systems within TRIZ-based innovative projects. They allow TRIZ practitioners to identify system disadvantages that are hard to reveal using the Classical Function Approach.

Finally, we introduced a novel approach to categorizing functions that engineering systems perform. This makes the Function Analysis procedure simpler and, at the same time, enhances the value of analytical results.

Acknowledgments We extend our sincere thanks to our colleagues—TRIZ developers who participated in discussions about function approach and importance of parameters for describing functions. Special thanks to Deborah Abramova who helped to make this paper sound more literate in English.

References

Ackoff, R. L. (1978). *The art of problem solving*. Hoboken, NJ: Wiley.

Chittaro, L., & Kumar, A. N. (1998). Reasoning about function and its application to engineering. *Artificial Intelligence in Engineering, 12*, 331–336.

Cummins, R. (1975). Functional analysis. *Journal of Philosophy, 72*, 741–765.

Fedosov, Y. (2013, August 1–3). *Statistics of "Elementary Functions"* (pp. 59–67). Proceedings of the International Conference TRIZfest-2013, Kiev, Ukraine, MATRIZ, St. Petersburg Polytechnic University Publishing House.

Feygenson, O. (2012, August 2–4). *Specifics of function analysis for engineering systems at the exploitation stage*. Paper presented at the MATRIZ Conference TRIZ-Fest 2012, at Lappeenranta University of Technology, Finland.

Feygenson, O., Lee, J.-Y., & Feygenson, N. (2014, September 4–6). *Novel approach to categorization of functions that engineering systems perform*. Proceedings of the 10th MATRIZ Conference TRIZfest-2014: Theories and applications, Prague, Czech Republic.

Gerasimov, V. M., & Litvin, S. S. (1991). *Basic statements of methodology for performing function-cost analysis: Methodological recommendations*. Moscow: Inform-FSA.

Hirtz, J. M., Stone, R. B., et al. (2002). A functional basis for engineering design: Reconciling and evolving previous efforts. *Research in Engineering Design, 13*(2), 65–82.

Houkes, W., & Vermaas, P. (2010). Technical functions. On the use and design of artefacts. In: P. Vermaas (Ed.), *Philosophy of engineering and technology* (Vol. 1). Dordrecht: Springer.

Ikovenko, S. (2006). *TRIZ Application for IP Strategies development*. Abstract of Master TRIZ thesis, Boston. Accessed January 20, 2015, from http://triz-summit.ru/ru/203864/204207/

Kaufman, J. J. (2009). *Value management: Creating competitive advantage*. Toronto, ON: Sakura House Publishing.

Kynin, A., & Priven, A. (2013, August 1–3). *Formulation of functions of an engineering system with the matrix of generalized parameters* (pp. 59–67). Proceedings of the International Conference TRIZfest-2013, Kiev, Ukraine, MATRIZ, St. Petersburg Polytechnic University Publishing House.

Litvin, S., Feygenson, N., & Feygenson, O. (2010, November). *Advanced function approach* (Vol. 9, pp. 92–102). 10th ETRIA World TRIZ Future Conference, Bergamo-Italy, Elsevier.

Macnamara, L. (2012). *People and permaculture: Caring and designing for ourselves, each other and the planet*. East Meon: Permanent Publications.

The Strategy Reader. (2004). Susan Segal-Horn, editor, The Open University, Great Britain.

Vermaas, P. E. (2010). Focusing philosophy of engineering: Analyses of technical functions and beyond. In I. van de Poel & D. E. Goldberg (Eds.), *Philosophy and engineering: An emerging agenda*. New York: Springer.

Vermaas, P. E. (2013). The coexistence of engineering meanings of function: Four responses and their methodological implications. *Artificial Intelligence for Engineering Design, Analysis and Manufacturing, 27*, 191–202.

Weber, E., & de Preester, H. (2005). Micro-explanations of laws. In R. Festa, A. Aliseda, & J. Peijnenburg (Eds.), *Structures in scientific cognition. Essays in debate with Theo Kuipers* (Vol. 2, pp. 177–186). Amsterdam: Rodopi.

Part III
Essay

TRIZ as a Primary Tool for Biomimetics

Julian Vincent

Abstract The interface between biology and technology is well visited but bereft of theory. TRIZ is well suited to bridging such an interface but similarly lacks coherent theory, although it can be shown to work. In the absence of theory and associated numerical models, a semantic model has been developed by means of an ontology. The differences between the rigidity of theoretical ontology and the fluidity of biological adaptation and change can be surmounted by the adoption of the dialectic and the foundation of the TRIZ Contradiction Matrix. Biological terms, principles and morphologies can be interpolated into the standard Parameters and Inventive Principles of the Matrix. An example—the Mexican blind cavefish which escapes predation by moving into a dark environment and relying on senses other than sight—illustrates the workings of the dialectic and the necessity to describe the biological adaptation with several Inventive Principles. Thus, the BioOntology has a feature of OTSM, Altshuller's partially developed second generation of TRIZ.

Keywords Ontology • Biological evolution • Biomimetics • Pareto set • TRIZ matrix

1 Introduction

Any biologist who wishes to study biomimetics has to learn some engineering in order to understand the gap between the two sciences. Very few engineers have attempted the reverse journey. One of these few was Professor JE (Jim) Gordon, a pioneering materials scientist, whose book (Gordon 1976) guided many careers. His account of biological materials is very practical: in particular, he emphasised the superiority of the toughness of biological materials. As a result of reading this book, in collaboration with one of Jim's research associates (George Jeronimidis), I set up a pioneering Centre for Biomimetics at the University of Reading. We were often asked, as consultants, to help solve engineering problems. Our evolved technique

J. Vincent (✉)
Department of Zoology, University of Oxford, Oxford, UK
e-mail: julian.vincent@zoo.ox.ac.uk

L. Chechurin (ed.), *Research and Practice on the Theory of Inventive Problem Solving (TRIZ)*, DOI 10.1007/978-3-319-31782-3_13

was for George to define the problem in simple and basic engineering or physics terms; then I would translate that definition into a biological context; see how biology might solve the problem (again, in basic terms; often only in energetics) and translate that solution into an engineering context. George would take that new concept and develop it into a practical engineering concept. But there had to be a way to shorten and improve this process. An article in a trade magazine alerted me to TRIZ.

2 TRIZ Solving Biological Problems

Darrell Mann (University of Bath) allied TRIZ to biology (Mann 1999), and together we concocted a practical for my zoology class. After a short introduction to the Contradiction Matrix, the students were given a number of imaginary problems in zoology that they had to solve using it (Vincent and Mann 2000). For instance, a tree had developed shiny smooth bark as a defence against caterpillars feeding on its leaves. The caterpillars were unable to move on to the next leaf for fear of falling off. How could they survive?

The problem includes grip or adhesion, which resolves into force or strength. These have to be increased without compromising energy consumption, weight, durability or mobility. Contradictions identified and investigated by the students included:

Force vs Use of Energy suggests:

Periodic Action—a mechanism resembling the ribbon teeth of a snail that would extend the length of the underside of the caterpillar generating a literal caterpillar track! Each tooth on the ribbon would be in contact with the tree periodically. In fact, the hooked ends of the caterpillar's feet, of which there can be five or more pairs, are very much like this, the track being regenerated each time the legs are swung forwards.

Another Dimension—give the caterpillar a sticky tongue, like that of a chameleon, which it can shoot out to stick on to a leaf then haul itself up. Caterpillars actually do this in reverse, dangling from the end of a silk line until they collide with another part of the plant. Alternatively the entire animal could jump from one part of the plant to another (a common solution amongst adult insects, though not many caterpillars jump). Admittedly these modes of movement are rather more random than a designer might like but at the same time probably require much less energy than a directed mechanism and do not require the sophisticated sensory and ranging mechanisms which the glue gun tongue would need.

Force vs Convenience of Use suggests:

Self-Service—use a freely available or waste product to perform other helpful functions. The proverbial stickiness of dung, which the caterpillar produces, could modify the glass-smooth surface and make it suitably rough again. Alternatively the caterpillar can lay down a silken trail, using the adhesive properties of the silk. This is a common mechanism, especially amongst social caterpillars such as the pine processionary moth that lay down a path from an overnight nest area to the feeding area on the plant.

Force vs Weight suggests:

Segmentation—similar ideas to the caterpillar track idea above.

Mechanical Vibration—some kind of pulsed gripping action. Perhaps resonance would allow the development of a higher force without expending too much energy. No known examples.

Force vs Speed suggests:

Mechanics Substitution—'replace a mechanical solution with a non-mechanical solution'—a specific trigger to break out of the 'grip harder/better' and think about the use of adhesion or some form of chemical grip or possibly using some form of vacuum solution.

Area vs Strength suggests:

Local Quality—suggests keeping the feet the same overall size but to incorporate lots of small localised protrusions. In fact many adult insects can walk on glassy surfaces using nothing stronger than van der Waals forces to keep them there. They have compliant hairs on their feet whose ends can conform to the surface and provide a surprisingly large area of contact. This has since been popularised as the Gecko effect. Since the contact area is related to the weight (volume) of the insect, larger insects would have disproportionately large feet.

The students found the session very interesting. Clearly TRIZ can be popular with zoologists, but it is rarely presented in a relevant manner. A paper exploring the TRIZ biology interface rather more formally followed (Vincent and Mann 2002) in which the biological and engineering possibilities were developed further, including evolution and speciation.

3 Problems in Developing a Bio-TRIZ

In 2000, I moved into the Department of Mechanical Engineering with the remit to work out how engineers can easily access the design tricks that organisms use. The obvious tool for this project was TRIZ. Several papers resulted (Vincent et al. 2005, 2006) but although they attracted interest, the work did not properly marry biology and engineering. Altshuller had already examined this area and showed that in many instances, an interesting 'bionic' effect was recognised only after it had been shown in physics or engineering (Altshuller 1999). The same is true today—the hysteria about drag reduction of sharkskin riblets is a case in point. The concept already existed in engineering (Walsh and Anders 1989), and it was this that enabled their true nature to be recognised on an organism. Altshuller believed that mammals and other 'higher' animals are too complex for ideas to be easily extracted and that palaeontology offers the only hope of introducing biological ideas to his creation (Altshuller 1988). Although he recognised that biological species represent the patents of nature, he appeared unaware that 'lower' animals and plants often represent simpler, and probably more useful, ways of doing functions that 'higher' organisms display. For instance, he recognised that elephants replace teeth as they are worn out but was unaware that molluscs (snails, limpets, etc.) do exactly the same but with their teeth built on to a ribbon (the radula) that is continually fed forwards to the mouth as the front teeth wear out.

However, there is a basic problem here to do with the definition and presentation of a problem. Biology is descriptive, open ended and full of surprises. Engineering is numerical, closed ended and (ideally) never surprising. The common ground is that both are explorations of how to solve problems and implement the solution; give an engineer a problem and she/he will solve it. But what happens if you present an organism to the investigator? Every organism is a solution to the general problems of survival, ever improving, ever changing and ever adapting. But there is a difficulty with the logic: although we might guess at it, we don't really know what particular problem the organism is solving. Put simply, the engineer can tell you the answer to the question 'What is 6 multiplied by 7?' but the biologist can't produce a unique question to which the answer is '42'. There are millions of questions to which the answer is '42'. '40 + 2', '39 + 3', '2 × 3 × (5 + 2)' and so on. At this point, we need a biologist who is at home with engineering. Clearly Altshuller did not have access to such a person.

We can still compare biology and engineering in terms of the ways problems are solved (Vincent et al. 2006). Collating data from the Internet and from published papers, we find that whereas technology relies on energy and raw materials for the solution of problems, biology relies on information and structure. Thus whereas we are wedded to fossil fuels and increasingly rare minerals, biology produces similar or better results by embodying 'smart' design in hierarchical structures, using wet composites and cellular structures. Obviously biological materials can't perform at the temperatures found in jet engines, but it has evolved ways of dealing with high temperatures: the bombardier beetle can cope with temperatures of around 100 °C in the spray it produces, yet the materials from which the beetle is made would be expected to have degraded at 60 °C (Aneshansley and Eisner 1969). In this case, the trick is probably a combination of thermodynamics (everything happens very quickly) and insulation (the bubbles of oxygen which power the spray are produced around the edge of the reaction vessel and can provide the necessary heat resistance). We also, by reading Altshuller's original descriptions, in Russian, of his Inventive Principles, condensed the Matrix into one that we called that we called *Pravila Reshenija Izobretatel'skih Zadach Modernizirovannye* (PRIZM) and presented it in both technological and biological forms (Vincent et al. 2006). This was our first (and, as it turned out, only) success at building a TRIZ bridge between biology and engineering. But we failed in our efforts to derive basic rules of biomimetics. I don't think this is the fault of TRIZ.

4 A BioOntology: Better Than Nothing?

It has to be possible to build a bridge between biology and engineering, and TRIZ has to be the best tool. TRIZ presents techniques which are basically semantic rather than parametric, and that is the way that ideas are translated from biology into engineering. Numbers come later in the process. But it is impossible to know where the good biomimetic ideas will come from. It is therefore necessary to investigate the whole of biology and translate the processes used by living organisms in their solution to problems into the categories developed by the TRIZniks. The working hypothesis is that the design rules inherent in TRIZ are universal and complete. This is a reasonable starting point, as has been stated many times, but one that should be open to modification if it becomes necessary. However, a database cannot support the complexity of connectivity between biological and technical systems. It results in a cascade of 'one-to-many' constructs that become unwieldy. A better method is an ontology. An ontology can express complex relationships; it does not need to be redefined every time it needs to be reorganised as does a database; there are many examples of biological ontologies, particularly in medicine. However, there are none which bridge disciplines in the way that TRIZ does, and even the few TRIZ ontologies confine themselves to engineering.

TRIZ begins by recognising technical systems, defined by Savransky (2000) as 'any artificial object within an infinite diversity of articles [things] regardless of its nature or degree of complexity' and technological process, which Savransky defines as 'any artificial single action or consequences of procedures to perform an activity with [the] assistance of a technical system or a natural object'. This follows a well-trodden path leading back to Aristotle (and probably earlier) who first outlined these ideas. Analogously the Basic Formal Ontology (BFO) recognises two main categories: continuants (things that are permanent) and occurrents (events that happen) (Smith 2012). Unfortunately, this formal (essentially theoretical) approach to ontology, which purports to describe biological events and organisms, does not allow its occurrents or continuants to change (Smith 2012), rendering it useless in a biological context.

In biology, the class of static things might include types of habitat, lists of organisms (individuals and, more conveniently, species) that occur there, edaphic and climatic factors, etc. But it would essentially be a definition of a snapshot. Nothing moves. Nothing changes. The second class—dynamic processes—would include physiology and biochemistry.

Biology rarely works like that. Ecosystems develop and regress. Species adapt, evolve, change and proliferate. Biochemical pathways probably tend to be less variable, although there are so many that the organism, by upgrading or downgrading their contribution, can change its overall physiological and behavioural repertoire. Therefore, especially in biology, not only are things changing all the time, but it is difficult to draw a distinction between a biological 'thing' and a biological process. Both are expressions of adaptation, change and survival within an environment that is frequently hostile (predators, parasites) and is being exploited by its occupants.

These changes, inherent in biological evolution and associated speciation, can be compared with, perhaps even described or defined by, the thesis-antithesis-synthesis of dialectic materialism that underpins the TRIZ contradiction matrix. It could generally be said that organisms do not evolve unless conditions change that are outside of the control of the individual or the species. Changes may be driven by climate, predation, disease and so on, not all of which need be detrimental, as when the climate 'improves' or a parasite becomes benign or the organism moves into a new environment, for whatever reason. Conditions simply change. The problem is that a process cannot necessarily control its trajectory if it is sufficiently unstable that it *can* change. Most biological processes have built-in feedback mechanisms that allow them to adapt and stabilise. Otherwise the process risks being useless because it will generate chaotic variation that can undermine the stability needed to sustain a population of interbreeding individuals. The ideal must be a process that is responsive to change and can be affected by an external influence which can provide overall negative or positive feedback. The process is therefore adaptive. Eventually there will be a cascade of influences, so that ultimately those changes are a function of random events in the rest of the universe.

In TRIZ the imposed change eliminates the antithesis and thus resolves the problem. However, in the classical dialectic, thesis and antithesis are interchangeable. Examination of the TRIZ contradiction matrix shows that this is true for many of the thesis-antithesis pairs, more so for the more generalised ones from further down the list of Parameters, since the Inventive Principles suggested for the synthesis of the dialectic are the same whichever Parameter of the pair has been identified as 'good' or 'bad'. This realisation is important for dialectic reasoning in biology, since the dialectic may often be driven by the consequences of *not* evolving a better process leading to decreased chances of survival—the selective advantage of natural selection. The assumption is that *any* evolutionary change can be treated as an example of the dialectic (Engels et al. 1964). A TRIZ-based analysis of biology might therefore be able not only to quantify but also to explain evolution in a materialist fashion. The problem may also have a hierarchical dimension. For instance biological altruism makes sense only at the level of a family group or a population, not at the level of an individual (Hamilton 1964a, b). So depending on the level of hierarchy, the answer can be different. Experience with TRIZ and biology, therefore, is that an ontology is far better than a database to bring biology and technology together, thus generating a tool for biomimetics (Vincent 2014), and that TRIZ exposes weaknesses in the current ontological approach to biology.

5 A Working BioOntology

In practical terms, which will be less open to dispute and totally open to modification, the textbook versions of the Parameters and Inventive Principles of the Conflict Matrix need to be expanded in order to make the interface between biology and technology clearer. I show this in a series of figures illustrating the main factors in the TRIZ description of an aspect of the behaviour of the Mexican blind cavefish (*Astyanax* sp.), which lives in cave pools. In an experiment (Sharma et al. 2009), these fish were put into dark or lit ponds, together with a related sighted fish the Mexican tetra (*Astyanax mexicanus*). Under dark conditions, both sighted and blind fish follow the walls of the cave, but blind fish swim more nearly parallel to the wall and swim faster than the sighted fish. By contrast, in the light, the sighted fish don't move much for long periods or move slowly around the centre of the tank distant from the walls. Thus, wall following is a shared, primitive trait used for exploration in the dark and can compensate for blindness. The blind fish have become better at following the wall in large part due to better sensing with the lateral line, the fish's flow sensor. The blind fish thus have an advantage in the dark and are less likely to be found by sighted predators. So although they have lost their sight (Parameter 24) they are safer (Parameter 30) (Fig. 1 a, b). The Inventive Principles are 2, 13, 22 and 23 (Fig. 2, c–f). In the figures I have shown, the current expansion I use for each of these Parameters and Principles, but only in sufficient detail to illustrate their use in this particular example. Each is three or four times more complex than shown here and is continually being modified and enlarged. Note that the left side of each of these expansions is conventional TRIZ, merging into biology on the right. Effectively this is the transition between biology and technology. Names in camelCase are also listed in a separate part of the ontology that is devoted to biological matters, so it is possible by following these leads to find out much more about the relevant biology. I have found this system to be relatively easy to use, and it makes much more sense biologically.

Note also that the Inventive Principles listed for this example do not represent alternative strategies for solving the problem as it would be in the technical use of the TRIZ contradiction matrix. Taken together, with the expansion of the Principles to include the biological factors, they describe the changes involved in this method of avoidance of predation in an integrated manner. In this respect, the description of the solution is much more akin to Altshuller's later development of TRIZ—that is, OTSM. In this approach, he wanted to describe solutions to problems much more closely but still using fairly general Principles. The increased accuracy comes from the overlap of several Inventive Principles. This is exactly what is necessary to describe the complex and integrated changes that biological organisms adopt when they undergo evolutionary change. However, it is necessary to refine the definition of the problem in a more accurate way; the most obvious is to increase the number of Parameters required to describe the problem. In the student class session reported above, this problem was overcome by considering a number or competing pairs of Parameters. This raises further issues that require resolution. Does the dialectic

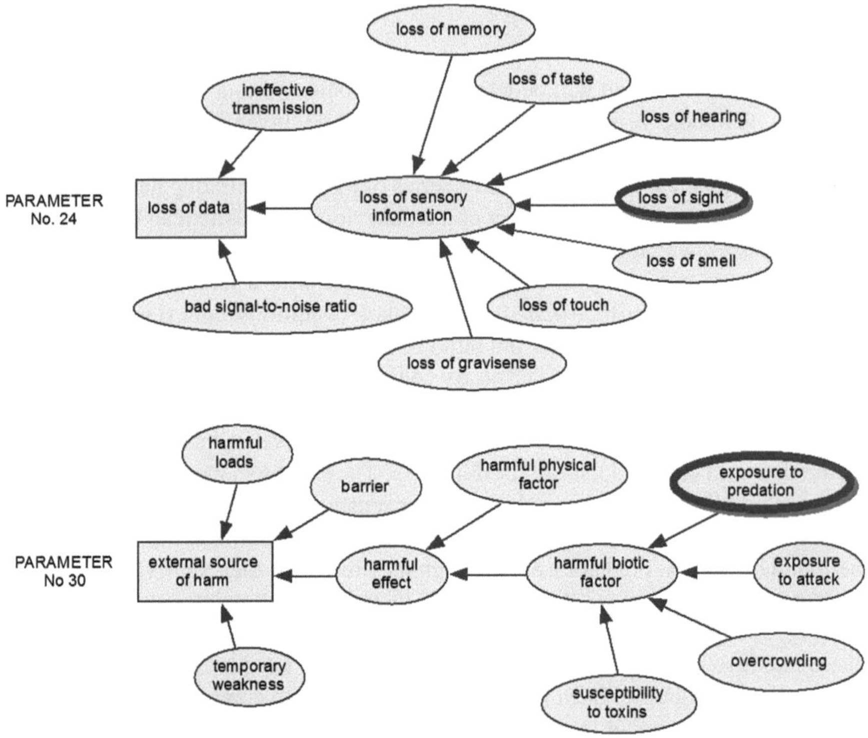

Fig. 1 (**a**, **b**) The Parameters describing the worsening (*top*, **a**) and improving (*bottom*, **b**) features of the problem of adaptation of the blind Mexican cavefish (16). The basic information about the Parameters was taken from Savransky (10), but all the biological references were generated during the creation of the BioOntology

confine itself to two dimensions (cf. the TRIZ matrix)? When analysing the research papers I use as my basic data for this ontology, it's not uncommon to find that there are several candidates for thesis and antithesis. But those two words allow for only two Parameters. Can we move into a 3D or 4D contradiction matrix? There is no reason why not. And if each biological problem demands three or four Inventive Principles for its description, there will be 40^3 (64,000) or 40^4 (2,560,000) different problem descriptions available, not taking into account the extra detail that the expansion of each Inventive Principle into the biological zone provides. TRIZ assumes that the dialectic presents a choice. Biology rejects that restriction in many instances. For instance, sexual reproduction involves the intermingling of products from two individuals, whether from the initial coupling or the growth of an embryo within the mother, which would normally excite the immune system. So for the purposes of reproduction the immune system of the female is partially deactivated, which leaves her more susceptible to disease. It's a necessary gamble. So reproduction and immunity emerge as physiologically incompatible and represent a dialectic pair. They are equally maintained, and the

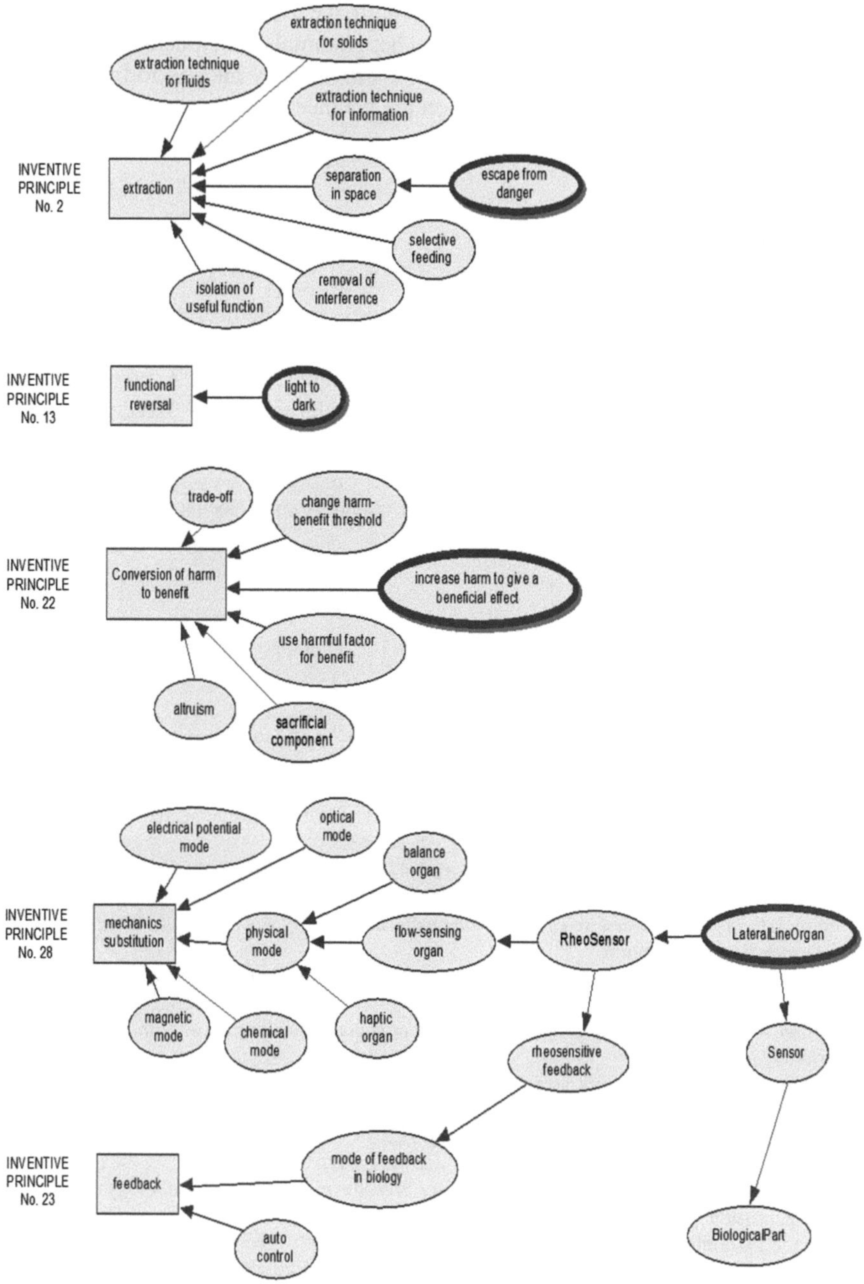

Fig. 2 **(c–f)** The Inventive Principles which the blind Mexican cavefish implements in order to escape from predators, thus resolving the problem posed in Fig. 1 a, b. This also shows how the lateral line organ is tied in with IP23 through its part in a feedback mechanism and is a biological part (a sense organ)

organism merely changes the balance between the two. There are many other examples of such a balanced system, in both animals and plants (McNamara et al. 2013). The analytical model required is provided by the Pareto front, which defines a range of outcomes depending on the balance between the dialectic pair.

6 Envoi

This brief overview of trying to integrate biology and engineering tells a story of experiment and phenomenology; the accumulation of information is indicative of the early stages of an area of study. There is no guarantee that TRIZ can be the main tool for integration, although the philosophical underpinning of TRIZ is compelling. The answer will come when we find a system that works.

Acknowledgements I am grateful to Brian C. R. Bertram, Norbert Hoeller, Filippo A. Salustri and Ashok K. Goel for helpful discussions

References

Altshuller, G. (1988). *Creativity as an exact science*. New York: Gordon and Breach.

Altshuller, G. (1999). *The innovation algorithm, TRIZ, systematic innovation and technical creativity*. Worcester, MA: Technical Innovation Center.

Aneshansley, D., & Eisner, T. (1969). Biochemistry at 100C: Explosive secretory discharge of bombardier beetles (*Brachinus*). *Science, 1656*, 61–63.

Engels, F., Dutt, C., & Haldane, J. B. S. (1964). *Dialectics of nature*. Moscow: Progress Publishers.

Gordon, J. E. (1976). *The new science of strong materials, or why you don't fall through the floor*. Harmondsworth: Penguin.

Hamilton, W. D. (1964a). The genetical evolution of social behaviour. *I. Journal of Theoretical Biology, 7*(1), 1–16.

Hamilton, W. D. (1964b). The genetical evolution of social behaviour. *II. Journal of Theoretical Biology, 7*(1), 17–52.

Mann, D. L. (1999). Creativity as an exact (biomimetic) science. http://www.triz-journal.com/archives/1999/11/c/index.htm

McNamara, K. B., Wedell, N., & Simmons, L. W. (2013). Experimental evolution reveals trade-offs between mating and immunity. *Biology Letters, 9*(4), 20130262.

Savransky, S. D. (2000). *Engineering of creativity: Introduction to TRIZ methodology of inventive problem solving*. Boca Raton, FL: CRC Press.

Sharma, S., Coombs, S., Patton, P., & de Perera, T. B. (2009). The function of wall-following behaviors in the Mexican blind cavefish and a sighted relative, the Mexican tetra (*Astyanax*). *Journal of Comparative Physiology A, 195*, 225–240.

Smith, B. (2012). Classifying processes: An essay in applied ontology. *Ratio, 25*, 463–488.

Vincent, J. F. V. (2014). An ontology of biomimetics. In A. K. Goel, D. A. McAdams, & R. B. Stone (Eds.), *Biologically inspired design: Computational methods and tools* (pp. 269–285). New York: Springer.

Vincent, J. F. V., Bogatyreva, O. A., Bogatyrev, N. R., Bowyer, A., & Pahl, A.-K. (2006). Biomimetics: Its practice and theory. *J R S Interface, 3*(9), 471–482.
Vincent, J. F. V., Bogatyreva, O., Pahl, A.-K., Bogatyrev, N., & Bowyer, A. (2005). Putting biology into TRIZ: A database of biological effects. *Creativity and Innovation Management, 14*, 66–72.
Vincent, J. F. V., & Mann, D. L. (2000, September). TRIZ in biology teaching. *TRIZ Journal.*
Vincent, J. F. V., & Mann, D. L. (2002). Systematic technology transfer from biology to engineering. *Philosophical Transactions of the Royal Society A, 360*, 159–174.
Walsh, M. J., & Anders, J. B. (1989). Riblet/LEBU research at NASA Langley. *Applied Scientific Research, 46*, 255–262.

Using TRIZ in the Social Sciences: Possibilities and Limitations

Joris Schut

Abstract TRIZ has gained an increasing presence in the engineering sciences in the last decades. However, due to its technical roots, its use in the social sciences is still very limited. Looking at three key areas of TRIZ, its philosophy, problem structuring tools, and idea generation tools, suggestions will be made how these areas could be adopted in social science in order to facilitate wider adoption of this method.

Keywords Social science • TRIZ

1 Introduction

Originating in electrical engineering, TRIZ has gained an increasing presence in the engineering sciences in the last decades. However, its use in the social sciences is still very limited. Given its potential to develop innovative solutions, a wider application of TRIZ outside engineering could help in solving complex social problems. This chapter will explore the possibility of using TRIZ to solve problems researchers, policy makers, and professionals face in the social sciences. This will be done by looking at three key areas of TRIZ: its philosophy, problem structuring tools, and idea generation tools. For these three areas, suggestions will be made how these could be adopted in social science.

2 Areas of Consideration

Due to its roots in the field of (electrical) engineering, using TRIZ in the social sciences is potentially difficult. Although it is believed the TRIZ philosophy is not limited to the technical sciences (Kaplan 1996), only few examples of this can be found. Examples in music, military strategy (Zlotin and Zusman 2006), the service

J. Schut (✉)
Department of Behavioural Science, University of Twente, Twente, the Netherlands
e-mail: jorisschut@hotmail.nl

L. Chechurin (ed.), *Research and Practice on the Theory of Inventive Problem Solving (TRIZ)*, DOI 10.1007/978-3-319-31782-3_14

industry [e.g., see Lin and Su (2006) or Zhang et al. (2005)], and the teaching of creativity (Fan 2010) show TRIZ can be applied in nonengineering environments.

However, these examples only show TRIZ can be used in these situations but do not provide guidelines to do so. This leaves practitioners in non-technical domains without guidelines to adopt this methodology. In addressing the question of how to adopt TRIZ in social sciences, a distinction can be made between three different components of TRIZ:

- The philosophy behind TRIZ (e.g., ideality)
- Tools related to problem structuring (e.g., Root Conflict Analysis, ARIZ, and ranking methods)
- Tools related to problem-solving (e.g., inventive standards and principles)

These three components of TRIZ will be explored in order to see if, and to what extent, TRIZ can be adopted in the social sciences.

3 Considerations Regarding the Philosophy Behind TRIZ

With regard to the philosophy behind TRIZ, the desire to find ideal solutions [that is solutions that have all the positives, none of the negatives, and achieve all the objectives (Mishra 2007)] is something that is also strived for in social sciences (Rubington and Weinberg 2011). An example of this aim for ideality in educational science is appreciative inquiry. The dream phase of this methodology lets users project an ideal situation to solve a problem. In the next steps of this method this ideal solution is transformed into an action plan to achieve this solution. A similar approach is taken in TRIZ which also aims to get as close as possible to a predefined ideal solution. Thus, the aims (reaching an ideal situation) are similar. However, appreciative inquiry focuses more on the progress toward ideality and TRIZ more on the development of (ideal) solutions. A second common point between TRIZ and social science methodologies is the acknowledgement solutions are perspective dependent. What might be ideal for one stakeholder, might be unacceptable to other stakeholders. This is acknowledged by both social sciences (Rubington and Weinberg 2011) and TRIZ (Mishra 2007).

As the goals behind social science and TRIZ are strongly similar, the use of TRIZ can be considered applicable for social problems from the perspective of the philosophy behind TRIZ.

4 Considerations Regarding Problem Structuring Tools

With regard to the tools that structure problems, the unraveling of complex social problems using the same principles as used in the engineering sciences can be questioned. In particular the possibility of expressing problems using only "IF" and

"AND" relationships in Root Conflict Analysis is questionable when working with interconnected social problems. "AND" relationships are most notable here, as they imply that solving one problem can lead to other (unsolved) problems becoming obsolete. However, the complexity of social problems is often targeted by a coordinated approach to solve all problems instead of just one. This would limit researchers and practitioners in the social sciences to only use "IF" relationships (where all underlying causes need to be solved instead of just one), instead of a combination of both ("IF" and "AND") types of relation. However, as problems are also dependent on each other, removing the "AND" relation might lead to situations that do not relate to the real situation. As there are only few examples of Root Conflict Analysis in the social sciences (e.g., see Schut, in press), more research is needed to understand better the use of "IF" and "AND" relationships in Root Conflict Analysis in the social sciences.

A second issue with Root Conflict Analysis is the causality it assumes between two connected elements. Although this is not a problem when it comes to the technical sciences, social science is much more cautious in the face of causality (Cliff 1983). Given that social sciences operate in systems that are not stable and continuously change (as opposed to controllable systems in the natural sciences), this is not surprising. Therefore, it should be noted that a method (such as Root Conflict Analysis) which implies causality will possibly not generate the most valid results when applied in social science. Although this does not mean the method is not useful in a social setting, it strongly implies the method has to be modified. These modifications will make the method more useful when used in the social sciences. A possible modification could be the use of correlations between two factors instead of a causal one. However, this has its limitations as well. Therefore, more research in this domain is needed as well.

5 Solution Generating Considerations

With regard to the tools that generate solutions, some limitations exist when using them in the social science. A good example of this is classes 2–4 of the inventive standards which are devoted to solutions requiring electromagnetic fields. As most of the classes of the inventive standards and principles were derived through patent research, these methods are primarily aimed at applications in a technical context instead of in the social sciences. However, adaptations were made to the original tools to fit business and organizational contexts by Zlotin and Zusman (1993a, b), and adaptations were successfully applied in both organizations and politics (Faer 1998). An example of this is: "To hit the competitor while not touching him" (positive effect, hitting the competitor; negative effect, not touching them) or "The absence of the enemy is a step toward defeat" (positive effect, absence of enemy; negative effect, defeat). When translated to a practical solution, this can become the following: finding someone who speaks negatively about the candidate who is not affiliated to your campaign (e.g., a disgruntled employee) or deciding not to take part in a public event (e.g., a specific election debate) and thus placing your competitor at a disadvantage. Also adaptations of the principles exist for journalism, marketing (Vikentiev 1992), and teaching TRIZ to individuals outside the engineering sciences [e.g., Fox (2008)]. Even though this illustrates that it is possible to "translate" the ideas from the technical sciences to other contexts, it also points out that there is a necessity to do so in order to generate the right solutions.

6 Conclusion

Based on the examples presented here, it can be concluded that, although it is possible to use TRIZ in the social sciences, some limitations exist that need to be addressed to allow a better use: ideality and the perspective it assumes can present a problem while dealing with many stakeholders, assuming that social problems use the same underlying principles as engineering is only partly true, and the highly technical solutions are not always applicable to social problems. However, with some modifications, these problems could possibly be resolved and would allow for a wider use of TRIZ.

References

Cliff, N. (1983). Some cautions concerning the application of causal modeling methods. *Multivariate Behaviour Research, 18*, 115–126.

Faer, S. (1998). *Methods of strategies and tactics of election campaign.* St. Petersburg: Stol'ny grad.

Fan, J. (2010). Application idea for TRIZ theory in innovation education. In *Proceedings of the 5th International on computer science & education* (pp. 1535–1540). Hefei: Computer Science and Education.

Fox, M. L. (2008). *Da Vinci and the 40 answers: A playbook for creativity and ideas.* Austin, TX: Wizard Academy Press.

Kaplan, S. (1996). *An introduction to TRIZ, the Russian theory of inventive problem solving.* Farmington Hills, MI: Ideation International.

Lin, C. S., & Su, C. T. (2006). An innovative way to create new services: Applying the TRIZ methodology. *Journal of the Chinese Institute of Industrial Engineers, 24*, 142–152.

Mishra, U. (2007). *Introduction to the concept of ideality in TRIZ.* Retrieved December 11, 2013, from http://papers.ssrn.com/sol3/papers.cfm?abstract_id=2273178

Rubington, E., & Weinberg, M. (2011). *The study of social problems* (7th ed.). New York: Oxford University press.

Schut, J. (in press). Recognition of prior learning in youth work in the European Union: Causes for its limited use and modeling the outcomes. *Italian Journal for Sociology of Education, 7* (1).

Vikentiev, I. (1992). Techniques of journalism: A bank of creative principles. *Journal of TRIZ*, 56–68

Zhang, J. Z., Tan, K. C., & Chai, K. H. (2005). A TRIZ-based method for new service design. *Journal of Service Research, 8*(1), 48–66.

Zlotin, B., & Zusman, A. (1993a). Creative models for business and management. *Journal of TRIZ, 93*(1), 82–91.

Zlotin, B., & Zusman, A. (1993b). Evolution of organizations and social systems. *Journal of TRIZ*, 73–81

Zlotin, B., & Zusman, A. (2006). Patterns of evolution: Recent findings on structure and origin. In *TRIZCONN2006* (pp. 1–49). Milwaukee, WI: Ideation International.

Linking TRIZ and Cross-Industry Innovation: Evidence from Practice

How TRIZ in the Context of Cross-Industry Innovation Can Turbocharge the Innovation Process

Peter Meckler

Abstract An innovation department of a medium-sized company without own research capacity needs to have 'their finger on the pulse of times', that is, to watch closely the specific fields of interest and to secure as early as possible the application of basic technologies by cooperating with external partners. One approach is to use some established creativity methods, which very often ends up in many ideas, which then are stored and never are regarded again. To overcome this stumbling block in the innovation process, a combination of TRIZ with open innovation is proposed. An evidence from practice is described by five selected examples, which have been quite successful. Even if these successes may not always be regarded as disruptive innovations, they show how quickly solutions without technical or physical contradictions can be achieved by combining internal and external sources.

Keywords TRIZ • Cross industry • Creativity • Examples from practice • Internal sources • External sources • Disruptive innovation

1 Introduction

Professor Ernst Pöppel from the Munich Institute of Medical Psychology (IMT) says, 'Anyone who wants to be creative has to outsmart his brain' (Pöppel 2005). In the course of time, many scientists racked their brains on how to optimally use the creative potential available in any individual. Whatever method we may use, we always depend on the support of other people. 'The best way to overcome one's personal limits is to join forces: Teams are more creative than individuals. In retrospect many good ideas seem to have been the accidental result of a sudden

P. Meckler (✉)
E-T-A Elektrotechnische Apparate GmbH, Industriestrasse 2-8, 90518 Altdorf/Nürnberg, Germany
e-mail: Peter.Meckler@e-t-a.de

L. Chechurin (ed.), *Research and Practice on the Theory of Inventive Problem Solving (TRIZ)*, DOI 10.1007/978-3-319-31782-3_15

flash of wit of a single genius. In reality, however, they are the result of hard team work, even in the case of professional creative workers' (Schmalholz 2005).

When *G. Altschuller* and *R. Shapiro* developed the method TRIZ ('теория решения изобретательских задач' = 'Teoria reshenija izobretatjelskich zadacz', approximately meaning 'Theory of inventive problem-solving') in the 1950s (Altschuller 2004), they thought at first of the systematics of technical apparatus and then deducted principles for inventing from more than 40,000 patents. They assumed that all major technical inventions were based on a few universally valid inventive principles which would allow solving physical and technical problems. The 40 inventive principles they found were confirmed over the years by other investigations.

The following essay does not so much deal with the TRIZ method as such, but will show by means of examples that this method may lead to amazingly innovative results in solution finding for technical problems in a very heterogeneous team and in a relatively short period of time.

2 Innovation by Gyro Gearloose Is Outdated

Inventing like Walt Disney's classic eccentric genius *Gyro Gearloose* did it, sitting alone in a laboratory, only surrounded by experimental setups and the neurons sending signals in his brain, is not the method any more for innovative companies today to be successful on a quick changing global marketplace. We need the inspiration of other people working on different disciplines. Only '...when you step into an intersection of fields, disciplines, or cultures, you can combine existing concepts into a large number of extraordinary new ideas' (Johansson 2006).

Meanwhile the disproportional growth rates in any technology field sometimes require not only cooperation with partners and research institutes but also with direct and indirect competitors—'coompetition'.

> Product life cycles, especially of high-tech products, were clearly reduced in the past years, and that means, less and less time remains to compensate the high innovation expenses by a sales success. For this reason, companies are encouraged in their own interest to better use existent know-how and ideas. The 'soft' factors of the innovation process like a good (knowledge) management and cooperation with external research institutes or suppliers for the purpose of transferring knowledge are receiving more and more emphasis (Institut der deutschen Wirtschaft Köln 2006).

This knowledge is not fundamentally new. Today, everybody knows in principle that innovation cycles are getting shorter and shorter. Starting from 1887, we have been content with 'scratchy' music from a record player for almost 100 years. From 1948 on, 20 songs had space on a disc of a diameter of 30 cm. In 1981, the compact disc (CD), which could store approx. 20 audio tracks on a 12 cm disc, was even being celebrated as something of a revolution. But less than 20 years later, the DVD was launched on the market—no difference in size, but with a multiple storage capacity for music and movies. Since 1998 memory sticks, a medium without moveable mechanical part, have become available. Today, 256 gigabytes are hidden on a Micro-SD card, the size of a shirt button. A 'Blu-ray Disc' with gigantic 400 gigabytes corresponding to a capacity of more than 90,000 MP3 music tracks is on the market since 2003.

While the CD could still be considered as a linear advanced development of the disc record, the memory cards and the new 'Super DVDs' are surely the innovation leap every technology freak dreams of.

Of course, many hundred scientists worldwide are always researching and developing such basic technologies in parallel. Nowadays, the results are immediately available via the Internet after publication for virtually every interested person. The use, however, has something to do with laws and patents and can only be realised successfully with a certain sense of the 'right things', a certain cleverness and the necessary analytic expertise. An innovation department of a medium-sized company without own research capacity therefore first needs to have 'their finger on the pulse of times', that is, to watch closely the specific fields of interest and to secure as early as possible the application of basic technologies by cooperating with external partners. In general, this is called 'technology scouting'.

Such research results of innovative technologies are then used by innovation departments to develop close-to-application prototypes, which are eventually finalised by R&D to products ready for production. This applies, of course, also to the innovative renewal of partial functions within an existing product in order to secure competitive advantages.

People like to call the staff of an innovation and technology department 'creative freaks', 'creative crackpots' or something similar. To a certain extent, they quite like these references and are even proud of being part of the department. However, to bask in the illusion that one is allowed to develop crazy ideas which may at some point result in a product is not enough in itself. Ideas have to be purposefully checked, protected and converted into working prototypes. Only then will the 'Skunk Works', as they are called at Lockheed (Rich and Janos 1994), obtain the company's internal necessary attention and acceptance, which are required to be efficiently innovative.

'All in all, German enterprises manage relatively well to get development ideas protected by patents... However, to achieve competitive advantages in the globalising competition, market-related patents and inventions also have to result in new products' (Institut der deutschen Wirtschaft Köln 2006). To secure this is one of the first tasks of an innovation department. For this purpose, however, processes and interfaces, especially between innovation department and product development, have to be defined in a way that handing the prototypes over to product development process takes place without any problems. Then, extraordinary ideas may result in extraordinary products which will be perfected by development teams during a linear development process. A development process from the idea to the final product is quasi-established. As in evolution, not all experiments lead to a perfect result, but elimination of failed experiments and further use of the positive interim results will create products which can occupy a singular market position.

3 TRIZ in the Context of Creativity Methods

At the beginning of every innovation, there is an idea. Every person knows places where he or she thinks to develop ideas best. 'Humans are linked to places by nature. For unfolding our potential we need safety and we feel safe when we feel at home in a place' (Pöppel 2005). This place can be the office, the balcony at home or the forest. But only in a team will the creativity slumbering within ourselves be loaded up so that it starts sparkling. Various methods were developed for this purpose. In his book James M. Higgins describes 100 methods of creativity and TRIZ (Higgins 2006), because TRIZ is not an actual method of creativity, but a system of purposefully using existing technical knowledge.[1] A clearly defined problem is at first transferred to an abstract level, and there solutions are sought by means of exactly such abstract innovation principles (in fact modern TRIZ is much deeper and wider than 40 inventive principles, but this basic aspect is the one we concentrated on). This very structured approach clearly differentiates TRIZ from all other described methods. A possible classification is shown in Fig. 1.

The methods shown in Fig. 1 are used during the various stages of the development process so as to fill the 'innovation funnel' (see, e.g. Brem 2007) with ideas or to generate concrete solutions related to the selected ideas. In this context TRIZ has to be considered a method for solution finding with technical challenges rather than a pure 'generator of ideas'. Nevertheless, TRIZ can be used all through the entire development process as a 'remover of mental blocks' and as a 'generator of solutions' (Savransky 2000). 'The time of individual innovation technologies can be assigned to the phases of the innovation process during which they are mostly used' (Koltze and Souchov 2011, pp. 3–5).

Some selected **creativity techniques** for generating ideas will be briefly presented in the following (see, e.g. Higgins 2006):

6-3-5 Six (**6**) participants collect **3** ideas each and get **5** min for each round. This method allows generating very many ideas in a very short time, although only a small fraction of them will be suitable for closer consideration.

Converse Question Often, there are mental blocks on the direct way to the target. Therefore, it might help to use the backdoor and do some brainstorming by putting the question the other way round. In the next step we look for the opposite.

WackPack If you want to go for something different—a pack of cards might give you suggestions for taking another point of view regarding the question or the solution. Like many other methods, this method also works with associations to create analogies for finding a solution.

Lotus Petal Brainstorming Around a blossom centre (the question/problem), the ideas for a solution are drawn as petals. The best idea will then be the starting point of another round of finding ideas.

[1] Position of the author.

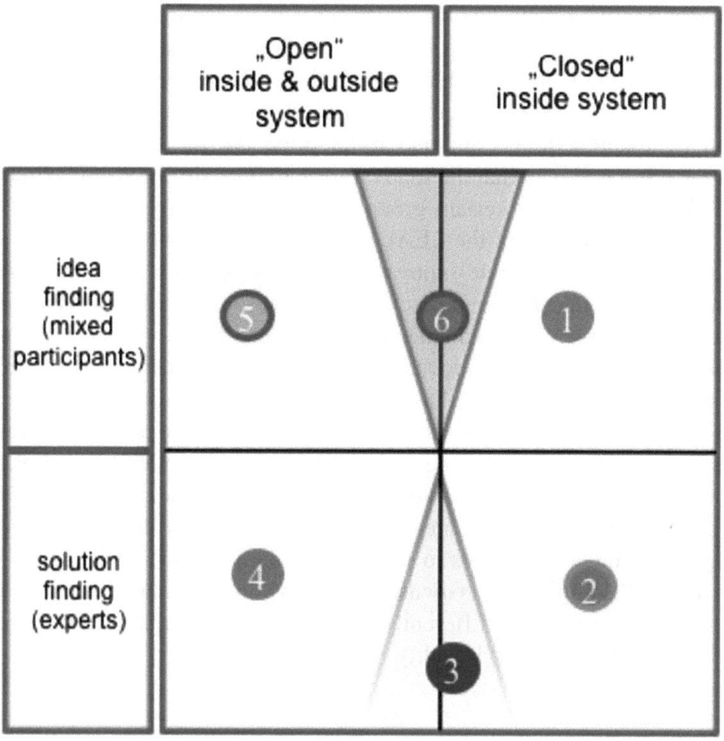

Fig. 1 Overview of methods. (1) Creativity techniques, (2) solution competitions, (3) networks (TRIZ), (4) technical alternatives (TRIZ), (5) lead user method, (6) internal crowdsourcing (see Sell and Meckler 2011)

This method is very popular in Japan.

Ten Faces We try to look at the problem from various points of view by adopting one of the ten typical innovator roles described by Tom Kelly. He distinguishes three categories: learning, organising and building roles (Kelley and Littman 2005).

Solution competitions at first are internal measures to obtain more solution approaches from different design engineers in the same time, and they are mostly used in the conception phase of the design process. However, solution competitions can also be carried out 'openly' and can then be assigned to the 'open innovation'. They can then be found on the left side of the method matrix in Fig. 1.

Building networks is a major aspect to build up an R&D department as a growing organisation which does not only 'stew in its own juice' but which searches flexibly and swiftly after technologies not only in its inherent sector but also in neighbouring sectors. The use of TRIZ has turned out to be an excellent tool in this zone between 'open' and 'closed innovation' to find **technical alternatives** to solutions moving in the same old groove, even if group activity and exercises itself are not the subject of TRIZ.

A major step in any product development is the selection of ideas after collection, e.g. with creativity methods. One way to find the right products for the right markets is the use of the **Lead User Method** developed by *Eric von Hippel* (von Hippel 1988). It is based on the empirical knowledge that innovations often are not initiated by manufacturers, but that users often are the driving power behind innovations. It has shown that the innovative activities are not spread haphazardly, but they concentrate on a certain group of particularly progressive users (LEAD users). It is the clear goal of the LEAD user method to involve these pioneers at a very early stage into concrete innovation projects of firms. Already in the pilot study stage of a product design project, the LEAD user method looks at additionally suitable approaches from analogue, i.e. from sectors of the same nature. The systematic development of innovation for tomorrow's markets and minimising the failure risk at the same time has top priority.

Free and easy discussions in casual surroundings can help to make withdrawn or shy design engineers communicate with their colleagues so that the company itself is able to 'tap' the wealth of creative potential existing in the company. This can be in the shape of an 'innovation breakfast' in the canteen or during an 'innovation meeting' with outdoor activities to shed off the daily routine (Figs. 2 and 3). This can also be called **internal crowdsourcing**, while building up inhomogeneous teams with participants of different departments of an organisation, e.g. R&D, finance, purchasing, sales and quality.

Fig. 2 Unusual team building exercises such as roping down from a rock can help to get rid of obstructive ballast of mind as body and mind will be completely absorbed by the task

Fig. 3 Creativity workshops should take place outside of the usual office surroundings, e.g. in a cave, drinking mulled wine that might light up the mind

3.1 How TRIZ and Cross-Industry Innovation Can Turbocharge the Innovation Process

'80 percent of all innovations are a re-combination of existing knowledge. [...] This means an innovation will mainly be generated when existing knowledge about markets, products, technologies, application principles or business models can be combined anew' (Enkel 2014). Opening up the innovation process towards the outside is a tendency no longer ignored by many major companies. 'The cross-innovation approach fits into the new "open innovation" principle which makes innovation a job of many players who can also come from outside the company' (Steinle 2010).

20,000 academic treatises are published globally each workday (nibis). Global increase of knowledge is snowballing and can no longer be handled by the individual. Using our 'swarm intelligence' will be a necessary precondition for developing successful and market-driven products.

So what could be more appropriate than combining the more inward-looking ability of TRIZ with an open method such as 'cross-industry'? According to (Enkel 2014), TRIZ helps with abstracting a problem and cross-industry helps with finding analogies, but they only look at a very narrow search field so that solutions may be lost. This may be right but is not so important when it comes to finding quick and consistent solutions. In addition the 'not-invented-here syndrome' will not fall onto any fertile ground in heterogeneous groups made up from different industries and different companies, and the internal competition will temporarily be skipped. Open innovation events will obviously push an innovative culture which is of major importance for the success of an innovation. 'Equally important for the success is a company culture which supports innovations and welcomes alterations. The will to work with colleagues and external partners in an interdisciplinary team will then quickly bear fruit' (Elmer 2013).

At the *Chair of Industrial Management of the Friedrich-Alexander University Erlangen-Nürnberg* (FAU), an innovation society called *quer.kraft—der Innovationsverein* was founded by Alexander Brem (see 2007) in 2008, which particularly deals with networking and the exchange in best-practice meetings about the topic of 'management of ideas and innovations' between companies in Southern Germany. During a best-practice meeting and more on a whim, one had the idea to establish a research group TRIZ. While it started up sluggishly with only three companies in the beginning, the group developed rapidly, based on quick successes, and became a 'self-runner'. Today, we have a growing number of approximately 15 participating companies, and we already had 10 successful TRIZ workshops about a wide range of topics.

It is exactly the heterogeneous composition of the group that yields results even amazing for the experts in the involved companies. The hosting company brings along a problem which is then presented to the team. In a one-day workshop, the problem is tackled with the typical TRIZ tools. These include:

- Ideality/ideal final result (IFR)
- Ideal machine
- Resource analysis
- 40 inventive principles
- Contradiction analysis
- Technical contradiction
- Physical contradiction
- Substance-field analysis
- Root conflict analysis (RCA)
- Small dwarfs

We do not want to describe the methods here; for details, please see the relevant technical literature. A detailed description of the TRIZ tools with examples can be found in, e.g. Koltze and Souchov (2011). A lot of information and material for downloading are available on the website of *Robert Adunka*, where the 40 inventive principles are illustrated with very fine examples (http://www.triz-consulting.de/hilfsmittel/). The matrix established by *Darrel Mann* (Mann 2009) has turned out to be basically indispensable for running a workshop if you want to resolve contradictions with the 40 inventive principles.

Some practical examples which were successfully treated in the TRIZ study group of *quer.kraft—the innovation society* will be briefly described in the following. My readers may forgive me that I cannot give a more detailed presentation for reasons of confidentiality. There is a gentlemen's agreement in place in the study group that all rights of the solutions found will stay with the company who posed the problem. This has been working out fine so far, not least because only

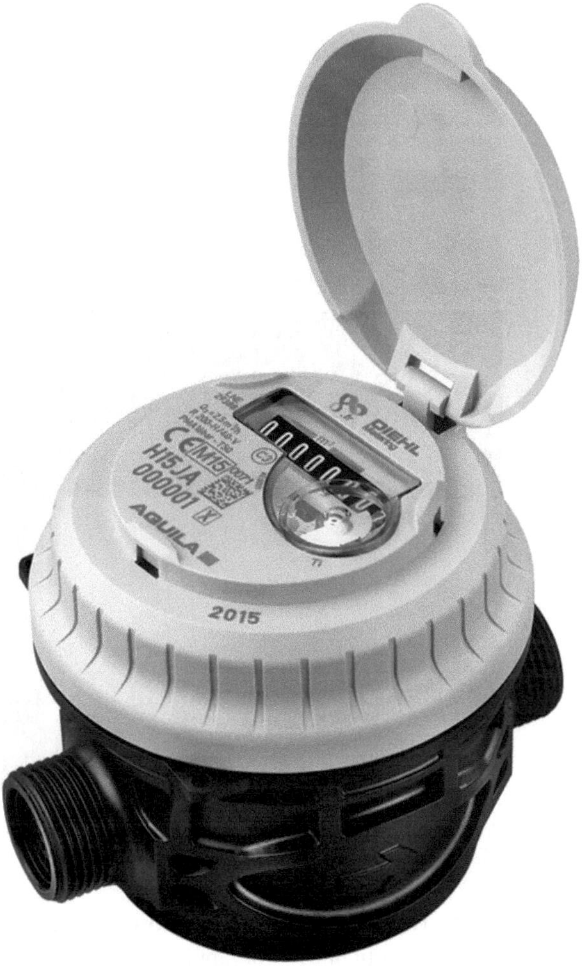

Fig. 4 Mechanical water flowmeter (photo: Diehl Metering GmbH)

non-competing companies from different industries take part. The product spectrum could not be more versatile and includes ignition plugs, engine pistons, cable harnesses, ball bearings, circuit breakers, roof windows, metres and precision motors. But it is exactly the 'non-experts' with a deep technical knowledge who are the 'turbochargers for disruptive innovations'.

Example 1: Mechanical Water Flowmeter Water condensation causes problems when trying to read out a mechanical water flowmeter or even makes it impossible. A particular challenge is that a comprehensive technical standard is in place on this score which must not be violated (avoiding a patent). Several solution approaches were established which had not been known yet and which are now implemented in the product. Feedback: 'Success story! Many thanks to the TRIZ workshop team' (Fig. 4).

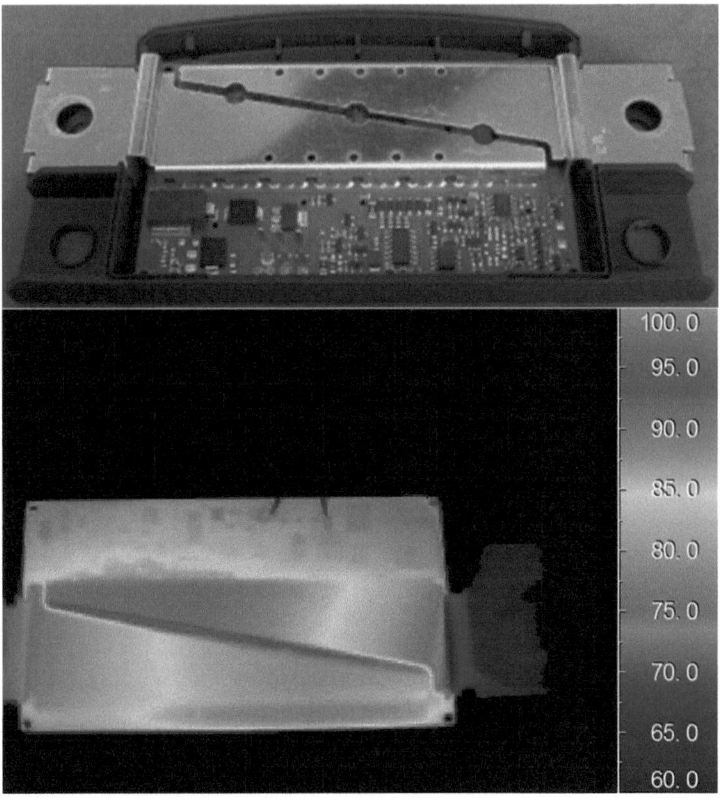

Fig. 5 Printed circuit board with high current connection (*left*, pcb; *right*, heat image at rated current) (*Source*: E-T-A GmbH, Altdorf)

Example 2: High Current Connection The high packing density on printed circuit boards, particularly for applications in the automotive industry, brings forward extreme requirements regarding the electronic components, but also regarding the design and connection technology, when the required currents are in a range of 50–300 A and the max. ambient temperature can be +85 °C. Contradictions between contacting high currents on pcbs, good heat dissipation and a flat design had to be solved.

14 innovative solution approaches were created, two of them led to patent applications (Fig. 5).

Example 3: No-Tool Handle Mounting with Safeguard on Roof Window Roof lights are supplied with dismounted handle to improve stackability and to avoid damages in transit and during unpacking on the construction site. Assembly should therefore be intuitive (poka-yoke), without tools ('plug-and-play') and protected against sliding out again (safeguard). A total of 34 individual ideas were generated with technical and physical contradictions and the dwarf method.

Fig. 6 Roof window with handle (*Source*: ROTO Dach- und Solartechnologie GmbH, Bad Mergentheim)

An implementation on an existing product is not planned, but the ideas will be stored for new developments and product enhancements (Fig. 6).

Example 4: Exact Positioning from a Standing Start—'Stick-Slip Phenomenon' When starting up from a standing start, drives/gearings in machine tools or robotics tend to have a poorly controllable behaviour (motor start-up/ 'jerking moment', stiffness, back lash, stick-slip phenomenon).

The group looked for an improved system allowing immediate sufficient control, predictability and thus exact positioning, if possible from a standing start. The problem analysis led to physical and technical contradictions which were solved by also using the 40 inventive TRIZ principles. The result of using this and other TRIZ tools were 111 (!) individual ideas. A patent application is considered (Fig. 7).

4 Outlook/Perspective

The ideas obtained by means of the line of action described above do by no means already represent an innovation. There is still a long way to go in the companies for the R&D departments. The dilemma of the innovator that well-established companies often refrain from investing in risky technologies, which might even be meant for markets not yet existing, is not solved with a disruptive solution (Christensen et al. 2011).

Fig. 7 Engine-gearing unit: TPM actuator made by Wittenstein (*Source*: Wittenstein AG, Igersheim)

The term 'innovation' is often used for reasons of market policy in order to emphasise the competitiveness of a company. Users mainly rely on the psychological effect of the word coming from the Latin verb *innovare* which means renew. Often innovation is only considered under the aspects of technology and products. Today's entrepreneurial view recommends an overall method of approach. The British journal The Economist wrote: 'Innovation is now recognized as the single most important ingredient in any modern economy' (Kelley and Littman 2005).

The definition supplied by J.A. Schumpeter (1997) is still considered a landmark, although it stems from 1911. Schumpeter basically distinguishes between **innovations** and **inventions**. For Schumpeter the latter is the implementation of a new idea up to the prototype status, while innovation also includes the introduction into the market. In other words: the invention only becomes an innovation with successful marketing. In this context an idea can concern a new product, a new production method, a new market area, a new source of supply or a new organisational structure. In practice it is often a combination of several alterations leading to the intended success. It should be mentioned in this context that modern TRIZ offers also special instruments for marketing analysis, including penetration.

The readiness to leave well-trodden paths is one of the basic requirements. 'The central competence to initiate and develop innovation strategies within a company is for us the readiness to look beyond the boundaries of one's own products, to look in the often unnoticed areas outside of one's own business fields and their interfaces' (Weisshaupt 2006). We have to be capable not to preclude a priori even the impossible. We have to seek the impossible everywhere, especially in places where so far we haven't been searching.

This is best illustrated by 'the story of a man who loses his key on his way home at night and who, despite his evident lack of success, restricts his search to the surroundings of a single street lantern. To the legitimate question of a policeman if he were sure to have lost the key in that exact spot, he answered: "No, I lost it back there, but there it is too dark"' (Weisshaupt 2006).

Enhancing the predetermined path of the TRIZ methodology and uniting it with the basic ideas of the cross-industry method have turned to be a turbocharger and have been the reason for achieving remarkable successes. Even if these successes may not always be regarded as disruptive innovations, they show how quickly solutions without technical or physical contradictions can be achieved by combining the internal and external aspects. The standard output expectation of a workshop described above is an incremental improvement in existing products, but disruptive ideas are always welcome!

References

Altschuller, G. (2004). *And suddenly the inventor appeared—TRIZ, the theory of inventive problem solving*. Worcester, MA: Technical Innovation Center. www.triz.org.

Brem, A. (2007). The boundaries of innovation and entrepreneurship—conceptual background and selected theoretical and empirical aspects, PhD thesis. Friedrich-Alexander-Universität Erlangen-Nürnberg (FAU), Fachbereich Wirtschaftswissenschaften.

Christensen, C. M., Matzler, M., & Friedrich von den Eichen, S. (2011). The innovator's Dilemma—Warum etablierte Unternehmen den Wettbewerb um bahnbrechende Innovationen verlieren. München: Verlag Franz Vahlen GmbH.

Elmer, E. (2013). Mit Cross Industry Innovation zu besseren Ergebnissen. *MM Maschinenmarkt, 34*, 22–23.

Enkel, E. (2014). Warum das Rad neu erfinden? Cross-Industry Innovation als neuer Trend im Innovationsmanagement, 3 M Die Erfinder—Blog für Innovation und Verantwortung. http://die-erfinder.3mdeutschland.de/innovationsprozesse/warum-das-rad-neu-erfinden-cross-industry-innovation-als-neuer-trend-im-innovati

Higgins, J. M. (2006). *101 Creative problem solving techniques—the handbook of new ideas for business*. Winter Park, FL: New Management Publishing.

von Hippel, E. (1988). *The sources of innovation*. New York: Oxford University Press.

Institut der deutschen Wirtschaft Köln. (2006). Wachstumsfaktor Innovation—Eine Analyse aus betriebs-, regional- und volkswirtschaftlicher Sicht. Deutscher Instituts-Verlag GmbH, pages 20 and 23.

Johansson, F. (2006). *The Medici effect—what elephants & epidemics can teach us about innovation* (p. 2). Boston: Harvard Business School Press.

Kelley, T., & Littman, J. (2005). The ten faces of innovation, IDEO's strategies for beating the devil's advocate & driving creativity throughout your organization, Doubleday, S. 210.

Koltze, K., & Souchov, V. (2011). *Systematische Innovation—TRIZ-Anwendung in der Produkt- und Prozessentwicklung*. München: Carl Hanser Verlag.

Mann, D. (2009). Matrix 2010—Re-update der TRIZ Widerspruchsmatrix, Copyright 2013 Naehler H. Th., Gronauer B., Gundlach C., c4pi—Center for Product Innovation http://wp.c4pi.de/ [Title of the English original edition Matrix 2010—Re-updating the TRIZ Contradiction Matrix, 2009 Copyright Darrel Mann].

Pöppel, E. (2005). Heureka, ich hab's gefunden, manager-magazin.de 2005. http://www.manager-magazin.de/unternehmen/karriere/a-341834.html

Rich, B. R., & Janos, L. (1994). *Skunk works—a personal memoir of my years at Lockheed*. Boston: Little, Brown and Company.

Schmalholz, C. G. (2005). Neuer Weg nach oben, manager-magazine.de 2005. http://www.manager-magazin.de/magazin/artikel/a-342110-7.html

Schumpeter, J. (1997). Theorie der wirtschaftlichen Entwicklung—Eine Untersuchung über Unternehmergewinn, Kapital, Kredit, Zins und den Konjunkturzyklus. Duncker & Humblot, Auflage: 9. A., unveränd. Nachdr. d. 4. A. 1934., S. 100 ff.

Savransky, S. D. (2000). *Engineering of creativity—introduction to TRIZ methodology of inventive problem solving*. Boca Raton, FL: CRC Press.

Sell, C., & Meckler, P. (2011, September 21). Einsatz von Kreativitätsmethoden an ausgewählten Praxisbeispielen, 2. Jahrestagung von quer.kraft—der innovationsverein, Nürnberg.

Steinle, A. (2010). Die Cross-Innovations-Methode, see the study: Praxis-Guide Cross-Innovations. Wettbewerbsvorteile durch einen branchenübergreifenden Innovationsansatz, Horx Zukunftsinstitut GmbH. http://www.horx.com/zukunftsforschung/Docs/02-M-16-Cross-Innovations.pdf

Weisshaupt, B. (2006). Systeminnovation—Die Welt neu entwerfen. Zürich: Orell Füssli Verlag AG, pages 20 and 109.

http://www.nibis.ni.schule.de/~igsm/medien/A2.htm

TRIZ and Big Systems

Bakhturin Dmitriy

Abstract To date, the practice of TRIZ community has accumulated a lot of experience with small- and medium-sized businesses, as well as in innovative projects of different scales. In this paper the specifics of application of TRIZ in the so-called big systems (primarily meaning big businesses, corporations) are analyzed. The focus is on the particularities of large systems that affect the practice of TRIZ, as well as on the problems and gaps in the theoretical basis of TRIZ which complicate its use in such systems. As a methodological framework, Hegelian triad known as "target," "object," and "means" (which is the minimum set necessary for meaningful activity) is used (see Hegel "Science of Logic," Vol. 2 Section 2 "Objectivity" Chapter 3, p.B "The Means," especially § 1607). Applying to TRIZ in general, the triad can be interpreted as "development" (target), "technical system" (object), and"TRIZ" (means).

Keywords TRIZ • TRIZ practice • Big corporations • Large-scale projects • Development • Evolution • Poly-systems • Man-machine system • Organizational and technical system • TRIZ elaboration

1 Specificity of TRIZ as a "Means" for Development of Big Systems

Successful application of TRIZ in companies such as Samsung, POSCO, and General Electric is well known in the TRIZ community. *What do we know about the functions of TRIZ there?* The best known and most frequently discussed is the experience of Samsung. As we can see from the reports of our colleagues, the company practices (a) advice to project teams and (b) corporate training at all levels. GE goes the same way. POSCO has declared a deeper, integrative approach arguing that TRIZ has become the part of the company's "ideology," a part of its mission statement. And while project TRIZ consulting in big systems is arranged

B. Dmitriy (✉)
Central Institute for Continuing Education & Trainings of State Corporation "ROSATOM", 21 Kurchatov st, 249031 Obninsk sity, Kaluga region, Russia
e-mail: rozmysl@mail.ru

L. Chechurin (ed.), *Research and Practice on the Theory of Inventive Problem Solving (TRIZ)*, DOI 10.1007/978-3-319-31782-3_16

virtually the same way as in small and medium businesses, it is general training and "ideology" which specifically characterize the modus vivendi of TRIZ in big systems. This seems logical if we take into account that one of the key challenges for big systems is to reduce the so-called transaction costs, the costs of organizing communication and interaction between large number of services, branches, and staff. TRIZ in this sense ensures the function of a general engineering language, attitudes/approaches, and perceptions, allowing the company to create a basis for effective development and implementation of "big projects."

Today a number of "methodologies" compete for the role of being a common language and a system of concepts in engineering, namely, "systems engineering," "Lean-Kaizen," "Six Sigma and Total Quality Management," etc. These "competitors" have positive—and very strong—sides, and they are more popular and widespread than TRIZ; even more, they actually try to absorb TRIZ, treating it as an internal part. In this regard we often hear judgments like: "But we (TRIZ) are special. Only we are capable of actually solving the problems whatever methodology is chosen; you cannot do without us." This is very much true, but only at the level of "project consulting," small and medium business, and outsourcing. For big systems at the "top level," this is not sufficient. Big systems are primarily interested in an integrated approach that can provide turnkey solutions. At the corporate level, as a rule, there can be only one "ideology" ("methodology"), and it is difficult, expensive, and time consuming to change one for another. If you have already implemented the "Six Sigma" or "Lean-Kaizen," this approach is going to be implemented "throughout." Under such circumstances TRIZ is forced to adapt the logic and the tools of the "leading" methodology, to "play by its rules," and to solve "its" problems. Thus, TRIZ methodology itself is reduced to a tool of the lowest level, with obvious loss of reputation and limited access to resources for its own development.

This is one of the challenges for TRIZ deployment in big systems—the ability to be a common "ideology" of engineering, to serve as a common language and conceptual "medium," and to "cover" all areas of engineering problems. In particular therefore, recurring discussions about the scientific nature of TRIZ are of a practical nature: "scientific" today means not only and not so much "reproducibility" and verity of the practical results obtained but also "logic," depth, and consistency of the theory allowing it to perform the above functions. In big systems these functions are even more important than the instances of individual "stellar" success of TRIZ consulting. Undoubtedly, development and competitiveness of the theoretical, methodological, and didactic components of TRIZ are the key aspects of its progress in big systems.

2 The Specificity of the Goals of Big Systems and TRIZ Model of Development

"Development," "growth," and "progress" are the traditional goals of TRIZ, internal part of it. It is commonly assumed that the "destiny," "success," and "survival" of companies depend on development. However, this is true only for small and medium businesses and even so not for all of them but only for "start-ups" and "technology companies." For big systems the complete opposite is true. The main aim of a big system is to "reproduce" itself, to prolong its own existence. Development of big systems is subject to reproduction; it is a way to ensure reproduction, and unless absolutely necessary it is not required. Metaphorically, we can describe a typical big system by such characteristics of technical systems of the third stage of evolution (using TRIZ "laws of technical systems evolution"): "limits of development are reached," "large number of small inventions," "focus on reducing costs and increasing service functions," and "in the long term it is necessary to change the principle of operation." And though this is just a metaphor, it should be admitted that, in relative terms, there is little "place for TRIZ" in big systems and that this topic is rather on the periphery than in the focus of attention.

No doubt, large-scale development is also present in big systems. It is realized through "big projects"—whether these are projects of modernization, restructuring, or making innovative products (which according to metaphors of the laws of technical systems, evolution is the equivalent to the transition from one S curve to another). As a rule, these are high-tech long-term expensive projects that involve a large number of participants. They certainly need and provide an opportunity for application of TRIZ but—and this is important—such projects are relatively few, and it is necessary to "enter them" at a very early stage, at the very beginning. As the project unfolds, it accumulates "the inertia of the decisions taken" which then cannot be broken even by TRIZ. This inertia is typical for all projects, but in big systems, due to their scale, it is too strong, and often even very effective inventive solutions proposed at later stages of the project are unable to outweigh it neither administratively nor economically.

From a theoretical perspective, when discussing the issues of "development," it is necessary to clarify the meaning of one of the key tools of TRIZ—"the laws of development of technical systems." In the Russian language and in the interpretation of classical TRIZ, we refer to "the laws of **development**." However, given the current practices and the role of TRIZ to provide real development, it is better to speak of the laws of **evolution** of technical systems. Conceptual distinction between "evolution" and "development" allows us to specify "development" as a certain function designed to "violate" the laws of evolution, for example, to accelerate the emergence of certain characteristics and the decline of others ("selection," if we adopt the logic of the "evolutionary approach"). The practical meaning of this distinction is to more clearly and rigidly define the specifics of development when using TRIZ. We can say that "the evolution of technical systems" is driven by standard engineering approaches within the standard organizational structures.

In contrast to “evolution,” “development of technical systems” is ensured by the use of “nonstandard” approaches, the most advanced of which is TRIZ. Metaphorically speaking, everyone evolves—only a few develop!

3 The Specifics of a Big System as an Object and Some Theoretical Issues of TRIZ

It should be admitted that at present, there are no specific TRIZ concepts of big systems. This can be illustrated by such TRIZ tool as a system operator. Unlike “simple systems” big systems are never “mono-process” and “mono-target.” It is impossible to find the “one in charge,” to describe “one past” or “one future,” and to specify “one super-system” or “one subsystem.” Any person, any object, and any product are included in a number of processes, programs, spheres of influence, and supersystems. Referring to Fig. 2, we see that, for example, the central square (“system in the present”) is a part of different “sets of systems”: SuperSystem1 + System + SubSystem1, SuperSystem2 + System + SubSystem3, SuperSystem3 + System + SubSystem1, etc. All these bundles are equivalent (imagine that SuperSystem1, 2, and 3 stand for the deputies of director general, e.g., finance, personnel, and production deputies, system is the factory, SubSystem1 the workers, SubSystem2 the maintenance, SubSystem3 the accountants). In addition, it should be noted that for a “simple” System Operator 1 (Fig. 1), its structure and transition rules are set, while the “complex” System Operator (Fig. 2) does not have such instrumentality, thus becoming too abstract and poorly suited for practical application.

In application to big systems, some difficulties also arise from the use of today’s TRIZ concept of the “technical system.” According to the definition in /2/, “any technical system is created to perform a set of useful functions, to achieve certain goals.” Thus, “practicability” and “artificial nature” are features of a technical system. However, many objects are barely covered by this definition. Take, for example, the destroyed nuclear reactor at the Fukushima nuclear power plant. This is definitely a “system” in a certain sense. But does it have a purpose? Was it created by someone? Obviously, it came under the influence of artificial, technical, and “natural” factors. One can find more examples of such “models” in the nuclear industry—abandoned uranium mines, long-term storage, spent nuclear fuel, etc. The number of such “techno-natural” or “natural-artificial” systems will only increase, and TRIZ needs to modify the existing concept of “technical system” in order to ensure its adequacy to objects of practice.

Any big system is always a “big man-machine system” (or, as it is said, a “big organizational and technical system”). As for this, TRIZ has no models and appropriate terminology. We still talk about the “iron TRIZ” and “mechanical TRIZ,” contrasting it to “the application of TRIZ in nontechnical areas.” However, if the technical system is defined as having a “target” and being “artificial,” then

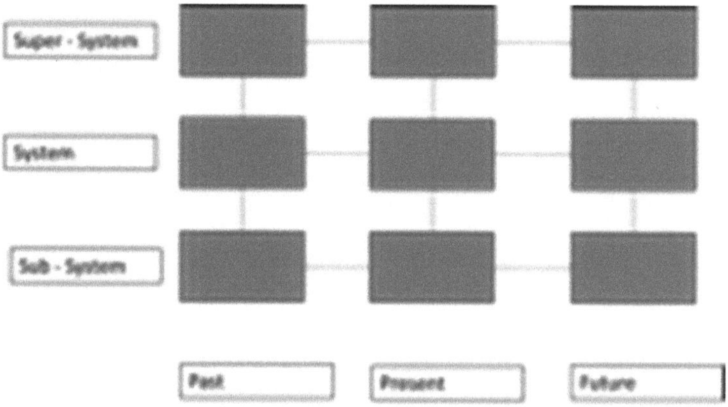

Fig. 1 System Operator 1

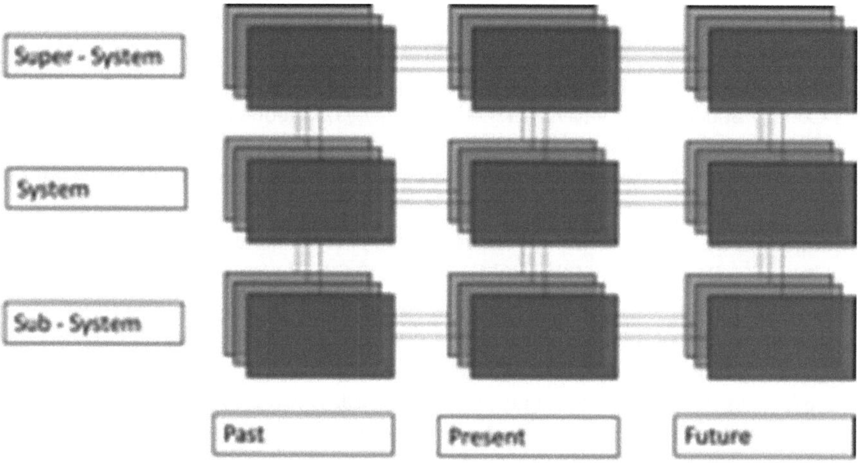

Fig. 2 System Operator 2

advertising and PR, elections, or any human collective undertaking (e.g., a shopping mall) may be considered as a "technical system." It seems that the resolution of this conflict is possible through extending the definition of "human" and "social," distinguishing it from the "iron" and "mechanical" in different interpretations of the technical system. For example, in the logic of the "iron/mechanical TRIZ," a nuclear power plant operator is a part of the "technical system" and cannot (should not) be considered as a person with their goals, attitudes, motives, and weakness (not to mention creativity). The operator is "only" a part of the "technical system," a part of the actuator, which has its own specific characteristics (e.g., speed of reaction). And—opposite to this—in the logic of social values, for example, even pure "hardware" suddenly gets a new feature and acquires a human dimension

Fig. 3 «Double» system

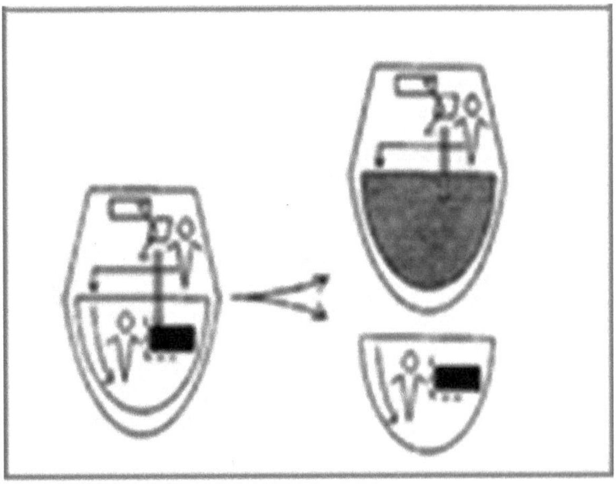

(e.g., when we convert a nuclear reactor into a museum like "The world's first nuclear power plant" in Obninsk, Russia).

Two schemes developed in the USSR-Russian methodological school [founded by Ceorgy Schedrovitsky (1929–1994)] (Fig. 3), /1/ can be given as an example, the "encompassing" and the "encompassed by" systems, where the inner (encompassed) hemisphere can symbolize human-machine operation and relation, while the outer (encompassing) hemisphere—the person-to-person relation—for example, is associated with the transfer of goals from the external system into the internal one. It must be said that publication /2/ on the "complete technical systems" suggests the same logic: as we know, the book states that "the majority of technical systems are not complete" (i.e., they are not purely "iron/mechanical" and always contain a person in a particular function). It is important to extend the definition of these human functions so that later, when necessary, it would be possible to clearly distinguish the "human" and the "machine" components of the inventive situation in order to implement proper tools. This fact becomes particularly significant for big systems when, as described above, a specific "mechanical" system is brought into the focus of responsibility of several different "human" systems.

Other significant difficulties should also be noted. TRIZ model of a technical system has no distinction between the "source material" and the "product" (as the result of the technical system altering the material). The existing definitions of technical system either do not specify the product at all or include it in the "final item" (in vepol analysis). This leads to difficulties in the analysis of long production chains and large and complex technological processes which, in fact, are typical of the production processes in big systems. Clearly, this observation is a technical one since availability of product and starting material is presupposed, but a number of tasks associated with, for example, the development of technology would be easier to discuss and solve if technical system models with dedicated states were available. Thus, for example, a connection could be made with such developing field as requirements management, where the requirements of the customer are the initial input giving certain starting and final states, and the actual engineering work is the work aimed at implementing those requirements.

Finally, TRIZ has no meaningful models of "supersystems" presently. For "ordinary" technical systems, considered in TRIZ—conveyor, device, product, etc.—big systems act as meta-systems. But what are the supersystems of the big systems themselves? Or of the large companies? It would be a mistake to say "the state," or "the country," because these are supersystems of a higher level. Probably we should be talking about "the industry" which in turn forms part of the "technosphere." But TRIZ has no models for such entities; more than that, there is no understanding of the laws of their development. In fact, besides the article by Altshuller "About the nature-absent technical world," TRIZ does not have its own understanding of the "technical world" and its laws of development. Hence the question: where do we take these concepts from for our own, TRIZ, practice and use? The situation with the concepts of social development is even more complicated. How does the society develop? What is the role of technology in this process? Are the old Marxian suggestions that "wealth is the amount of free time" and that, accordingly, "technology brings freedom to humanity" correct?

4 Summary and Recommendations for Further Work

- In order to advance at the level of big systems, it is necessary to intensify the theoretical, primarily methodological, and didactic development of TRIZ and to ensure coverage of all major classes of engineering problems.
- The primary method and direction for TRIZ to "enter" into big systems is penetration "from above," that is, through the upper levels of management, corporate programs, standards, policies, universities, business schools, and so on. Movement "from below" is doomed to the fate of the "poor inventor."
- The distinction between "evolution" and "development" must be clearly determined and fixed. Among other reasons, this should be done in order to emphasize the specificity of TRIZ in relation to other methodologies and the "standard engineering practice." In addition to theoretical benefits, this step can yield practical results facilitating interaction with the "regular" structures "engaged in development" within big systems.
- One of the main directions of development of theoretical and methodological parts of TRIZ for application in big systems is the development of concepts (notions) of "system," "technical system," "organizational and technical system," and "man-machine system," as well as harmonizing these concepts with modern ideas about the development of man, technology, and society as a whole.

References

/1/ Methodological school of management. (2014). Bloomsbury. ISBN, p 48
/2/ Altshuller, G., Zlotin, B. L., Zusman, A. V., & Filatov. V. I. (1989). Search of new ideas—from afflatus to technology. Kishinev: Kartya Moldovenska [on Russian language].

A Glossary of Essential TRIZ Terms

Valeri Souchkov

Abstract As any discipline evolves its body of knowledge, it might introduce and develop its own terminology to present key concepts and models. Since inception of TRIZ in the 1950s, a number of terms have been introduced to identify unique TRIZ notions and entitle tools which emerged on the basis of theoretical findings within the TRIZ studies. It is also important to note that several widely used terms which are known outside TRIZ have their own, more narrow, and specific meaning when used within the context of TRIZ, and it is therefore crucial to know such meanings to avoid misunderstanding. The chapter presents a collection of most essential terms which are used in the TRIZ body of knowledge and provides their definitions as well as comments and examples when necessary.

Keywords TRIZ • TRIZ terms • TRIZ glossary

1 Introduction

Since inception in the middle of the twentieth century, TRIZ (the Theory of Inventive Problem Solving) has remained a live discipline continuously evolved by numerous TRIZ developers worldwide. Modern TRIZ is a large body of knowledge which includes several dimensions: from a theory which attempts to study, model, and formalize the processes of inventive problem solving and technology forecast to a set of highly practical tools which use theoretical background and are designed to support engineers and technology professionals who aim at performing various innovation tasks.

Such broad development resulted in the emergence of a rather large extensive TRIZ-specific terminology which has been used within the TRIZ community to ensure effective communication. Some terms used within TRIZ are not new (e.g., such as the term "contradiction" or "function") but are used within specific context of systematic innovation and inventive problem solving, and therefore the meaning

V. Souchkov (✉)
ICG Training and Consulting, Willem-Alexanderstraat 6, 7511 Enschede, KH, The Netherlands
e-mail: valeri@xtriz.com

L. Chechurin (ed.), *Research and Practice on the Theory of Inventive Problem Solving (TRIZ)*, DOI 10.1007/978-3-319-31782-3_17

of each of such terms might have a slightly different, a more narrow, and specific interpretation than commonly known, and it is important to recognize such a difference.

The glossary of the essential TRIZ terms presented below contains a collection of such most broadly used terms of the Theory of Inventive Problem Solving. The terms included were selected from Souchkov (2014) and represent a selection of key TRIZ concepts and definitions. During the process of developing the glossary, a number of sources which introduced and discussed these terms were carefully examined (Altshuller 1998, 1984, 1993; Altshuller et al. 1989; Altshullers's Foundation 2013; Gorin 1973; Khomenko 2013; Litvin et al. 2013; Litvin and Lyobomirskiy 2013; Litvin et al. 2007; Matvienko 2013; Korolyov 2013; Petrov 2006, 2003; Petrov et al. 2013; Petrov 2002; Souchkov et al. 1991; Salamatov 1999; Zlotin et al. 1999; Zlotin and Zusman 2001; Official G.S. Altshuller foundation 2013).

The glossary consists of four parts, each focusing on a broad category:

General TRIZ Terms, which are used independently of their area of application within modern TRIZ.

Problem Analysis and Solving Terms, which are specific to solving inventive problems with TRIZ.

Terms related to the TRIZ tools, which are used to title TRIZ problem solving techniques and their core components.

Terms related to the TRIZ Theory of Technical Systems Evolution, which have a broader meaning than in the applications limited to the area of inventive problem solving only.

In each chapter, the terms are presented in alphabetical order.

Each term includes (or might include, depending on a necessity) the following fields:

Term: either a unique TRIZ term of a commonly used term which is used within special context in TRIZ.

Abbreviation: a commonly used abbreviation of a term.

Synonyms or alternative translations: In a number of cases, a term might have several synonyms which are used in the TRIZ-related publications or has several alternative translations from Russian. Several terms whose alternative translations are still widely used (e.g., "technical contradiction" and "engineering contradiction") were included as well.

Definition of a term.

Comment(s) to clarify the use of the term.

Example(s): In some cases examples are provided to better understand the meaning of a term.

The underlined words and expressions in the glossary mean that their definitions are available in the glossary.

2 General TRIZ Terms

Degree of Ideality
It is a dimensionless measure of an inventive solution, or a technical system, or a process, which identifies the degree of efficiency of the solution, the system, or the process through qualitative estimation of the ratio between useful functionality provided by the system/process/solution and a sum of costs to produce, maintain, and utilize the useful functionality.

Comment: The degree of ideality is primarily used to evaluate if a technical system/process/solution being analyzed is more ideal than a competing system/process/solution that provides the same main useful function.

Harmful Machine
Synonym(s): Harmful technical system

It is a model of a technical system which results from the process of extracting and understanding processes in the system which create negative effects.

Comment: The concept of a harmful machine helps with better understanding how problems featured by negative or harmful results emerge which simplifies their further elimination.

Ideal Technical System
It is a technical system that has an infinite value. For example, it may have neither components nor associated costs but still deliver the intended functionality.

Comment: Similar to the Ideal Final Result, such system may not exist, but its definition serves as a target to design the technical system with the highest degree of ideality possible.

Ideal Final Result
Abbreviation: IFR

It is a formulation of a final solution, which delivers the result required without adding neither extra material nor energy resources nor associated costs.

Comment: As follows from the laws of physics, such a solution may never be achieved, and therefore the concept of the Ideal Final Result serves to reduce the degree of psychological inertia during the problem solving process by targeting a problem solver toward searching for a solution with the best ideality ratio.

Example: *IFR for eliminating the heat produced by a microprocessor in a notebook PC can be formulated as "No heat is produced by the notebook PC without adding any material or energy resources to the system."*

Ideality
It is a dimensionless measure of an inventive solution which qualitatively identifies how closely the sum of compensation factors produce, maintain, and utilize the solution approaches zero value.

Innovative Task
It is a specific category of problem solving tasks which states that in order for a goal to be achieved, an inventive solution has to be found and implemented since none of known solutions may be used for one or another reason.

Examples: (1) To increase performance of a machine by 50 %, (2) to cut costs of a manufacturing process by 20 %, (3) to completely eliminate noise produced by the airplane engine in the airplane's cabin.

Inventive Problem
It is a situation which requires to perform a certain action either to synthesize a new technical system to deliver a new function, or to improve the function delivery by an existing technical system, or to prevent the technical system or its product from harmful internal or external factors in the situation when all known solution methods cannot be applied to achieve the result required.

Comment: The same inventive problem can be presented by different inventive problem models.

Inventive Problem Model
A model which only includes those components that are essential for further solving the problem with a specific TRIZ problem solving tool. In TRIZ, inventive problems can be modeled as technical, physical, or engineering contradictions, inefficient or harmful Su-Fields, inefficient or harmful functions, etc. A model that represents an inventive problem often is used as a part of inventive problem definition.

Inventive Problem Solving
It is a process which consists of a number of stages to find an inventive solution to an inventive problem.

Inventive Situation
It is a situation which is featured by a presence of a need to satisfy a specific demand of a supersystem without either a clearly defined specific problem to solve or a problem solving direction to be selected.

Inventive Solution
It is a solution to a specific inventive problem which meets all problem-specific demands and requirements and matches general criteria of an invention.

Level of Invention
It is a dimensionless qualitative measure which evaluates an inventive solution according to an estimated number of trials necessary to produce such the solution and the degree of its contribution to the general evolution of technology and engineering.

Comment: Currently five levels of inventions are known: (1) "Non-inventive invention" very simple invention that does not produce any significant impact on the evolution of a technical system; (2) invention that emerges from resolving a technical contradiction by a method available in the narrow engineering domain where the invention belongs to; (3) invention that emerges from resolving a complex technical contradiction by a method known in the engineering domain given; (4) invention that emerges from resolving a technical contradiction by a method available in a different engineering domain; and (5) pioneering invention that deals with complex inventive situations and which launches a radically new technology area.

Multiscreen Diagram

Synonym(s): (1) System operator; (2) Multiscreen of talented thinking; (3) Nine windows; (4) Nine boxes; (5) Nine screens; (6) Multiscreen scheme of talented thinking.

It is a method of analyzing a technical system which considers properties and features of the system in their relationships with subsystems and supersystem, as well as with previous generations of the system, its subsystems and supersystem, and their projections to the future generations of the system, its supersystem, and subsystems.

Operational Principle

Synonym(s): (1) Basic principle; (2) Working principle; (3) Principle of operation.

It is an abstract concept presenting how a required function can be obtained on the basis of a specific scientific effect or a combination of scientific effects and phenomena.

Comment: Any operational principle can be used to build a multitude of different designs of technical systems or their components.

Example: A basic principle of thermal expansion can be used in various applications and domains to expand and contract objects.

Subsystem

It is a component or a set of interacting components limited by specific borders that belong to a technical system. Both a separate material object and a larger system's part combining several components (material objects) can be regarded as a subsystem.

Example: A driver's chair, a steering wheel, and an engine are the subsystems of the technical system "car."

System, Technical

Synonym(s): (1) Engineering system; (2) Technological system.

It is a number of components (material objects) that were consciously combined together by establishing specific interactions between the components to deliver functionality with certain characteristics which are not possible to achieve by the components independently.

Comment: A technical system is designed, developed, manufactured, and assigned to perform a controllable main useful function or a number of functions within a particular context under specific conditions.

Comment: A technical system can include subsystems which can be considered as separate independent technical systems.

Supersystem
Synonym(s): (1) Supersystem; (2) A higher system.

It is a system that includes a technical system given as its part (subsystem). When using a concept of supersystem in problem modeling, a supersystem is limited to those components which interact or might potentially interact with the components of a technical system.

Example: A driver, cargo, a road, and an air outside a car are all components of a supersystem of a system "car." A bird is a part of a supersystem too since it can potentially interact with the car.

Systematic Innovation
It is a framework which combines various theories, methods, and tools based on a systematic approach to support innovative process from the initial situation analysis toward developing patentable solutions.

Systematic Technology Evolution
It is a hypothesis which states that evolution of majority of technical systems is governed by the same principles and patterns, irrespectively, an engineering domain or a technology area. Systematic technology evolution includes models of technical systems evolution (such as S curve and bell curve of evolution) and a collection of more specific trends and lines of technical systems evolution.

Theory of Inventive Problem Solving
Abbreviation: TRIZ

Synonym(s): Theory of Solving Inventive Problems

It is a scientific and applied discipline which comprises studying directions of technical systems evolution and a process of inventive problem solving in order to develop methods and tools to support innovative improvement of the technical systems based on a systematic and knowledge-based approach.

Comment: Lately, the TRIZ studies and developments have been extended to non-technology areas, such as business, social, and other types of artificial systems.

Comment: In the first translations from Russian to English, TRIZ was translated as TIPS. (2) The abbreviation "TRIZ" is the Russian acronym for the "Theory of Inventive Problem Solving" (Origin: "Теория Решения Изобретательских Задач" or "Teorya Reshenya Izobretatelskyh Zadatch", in Latin characters)

TRIZ Knowledge Base
Synonym(s): TRIZ knowledge bank

It is a database which contains high-level patterns and heuristics supporting inventive problem solving discovered as a result of research performed in TRIZ.

TRIZ Tool *Synonym(s):* TRIZ technique

It is a tool (technique) based on a certain modeling approach which describes a number of procedures to support one or another phase of inventive problem solving process or TRIZ-based evolution forecast.

Comment: Some TRIZ tools include databases of abstract solution patterns.

Examples of the TRIZ tools are the contradiction matrix and the system of 40 inventive principles, the system of 76 inventive standards, Algorithm of Inventive Problem Solving, and catalogues of effects.

3 Problem Modeling and Solving with TRIZ

Conflict, Typical
Synonym(s): (1) Typical contradiction; (2) Standard contradiction.

It is a type of a generic abstract problem model represented by either a technical or a physical contradiction which is frequently met in the models of diverse inventive problems and which in most cases can be eliminated by applying the same method of contradiction elimination despite a technology domain where the inventive problem belongs.

Example: A very common type of a typical conflict emerges when two objects must be in a direct contact with each other, but it leads to emergence of a negative effect.

Contradiction
It is a situation that emerges when two opposite demands have to be met in order to provide the result required. A contradiction is argued to be a major obstacle to solve an inventive problem and is used as an abstract inventive problem model in a number of TRIZ tools.

Comment: Three types of contradictions are known in TRIZ: (1) administrative, (2) engineering, and (3) physical.

Contradiction, Administrative
It is a description of either a negative (undesired) effect or a necessity to develop something new in a situation when neither a problem solving method nor a ready to use solution is available.

Comment: Although an administrative contradiction might not look as a contradiction since it misses a conflict between parameters or requirements, it indicates a conflict between the necessity to achieve the goal desired and available means to do it. In most cases, a formulation of administrative contradiction is similar to a formulation of an inventive problem without the use of any specific TRIZ-based model of an inventive problem.

Example: A car must be made safer without reducing its performance and increasing production costs, but it is not clear how.

Contradiction, Physical
Synonym(s): (1) Physical conflict; (2) Sharpened contradiction; (3) Contradiction of properties.

It is a situation that emerges when a certain attribute of a material object (represented as a substance or a field) must have two different (or opposite) values at the same time to provide a result required. An attribute can be a physical parameter, aggregate state, location, etc.

Examples: (1) Intensity of light emitted by the car's headlights must be high to let the driver clearly see distant objects in the dark conditions, and at the same time, intensity should be low to avoid blinding drivers in the oncoming cars. (2) A blade of a kitchen knife must be sharp to effectively cut food and at the same time should be blunt to avoid accidental cutting of a person's finger.

Contradiction, Technical
Synonym(s): Engineering contradiction

It is a situation which emerges when an attempt to solve an inventive problem by improving a certain attribute (parameter) of a technical system leads to unacceptable degradation of another attribute (parameter) of the same system.

Examples: (1) Improving the strength (one parameter) of the airplane wing leads to the increased weight (another parameter) of its wing. (2) Increasing the dimensions of a smartphone increases its battery capacity (one parameter) but leads to the inconvenience of use (another parameter).

Contradiction Elimination
Synonym(s): Conflict elimination

It is a type of a solution to an inventive problem which eliminates influence of one parameter on another parameter by decoupling the contradicting demands instead of applying parametric optimization, compromise, or trade-off.

Example: Instead of finding an optimal solution to a problem of how much energy to use to illuminate the surface of a the road sign at night, the surface of the road sign is covered with many tiny mirror segments which reflect the light emitted by the headlights of an approaching car.

Evolution of Su-Field Systems
Synonym(s): Evolution of substance-field systems

It is a hypothesis which states that elementary Su-fields tend to evolve over the time in order to increase their performance, quality, and other parameters. A group of problem solving methods structured in accord with the evolution of Su-field systems is collected and presented in the system of 76 inventive standards.

Field
It is a material object without rest mass that transmits interaction between components (subsystems) of a technical system represented as substances.

Comment: It is important to note that definition of a "field" in TRIZ differs from the definition of a field formally accepted in physics. Various terms presenting different types of energy exchange can be used. Examples of fields in TRIZ include the following fields: mechanical, acoustic, thermal, magnetic, electric, and electromagnetic. Sometime additional fields like intermolecular, biological, and informational are added.

Function

It is the specification of an action performed by a material object (function carrier) that results in a change or preservation of a value of an attribute of another material object (object of the function). In TRIZ modeling tools, a function is always represented as a triad: "a function carrier"-"action"-"object of the function."

Examples: (1) Flame heats glass; (2) a hammer hits a nail; and (3) a copper wire transmits electrical current.

Function, Harmful

Synonym(s): Negative function

It is a function performed by a material object (function carrier) that results in inacceptable change or inacceptable preservation of value of an attribute (parameter) or a state of another material object.

Function, Main Useful

Synonym(s): Main function

It is a physical action performed by a technical system which targets a certain object of its supersystem and that results in a positive (required) change or preservation of a value of a parameter or a state of an object of the function.

Comment: A main useful function has top priority in the technical system's functionality and is delivered to realize a purpose for which the technical system was designed within the context of its supersystem.

Example: Although a coffee machine can be used to heat water only, its main function is to produce coffee drink.

Function, Useful

Synonym(s): Positive function

It is a physical action performed by an object (function carrier) that results in a positive (required) change or preservation of a value of a parameter or a state of an object of the function.

Function Model

It is a model of a technical system resulting from function analysis that identifies and describes functional relationships between the subsystems of the technical system as well as between its subsystems and its supersystem.

Comment: Functions representing the functional relationships are characterized by category (useful, harmful, neutral), quality of functional performance (insufficient, excessive), cost level (insignificant, acceptable, and unacceptable), and cost of corresponding components.

Mini-Problem

It is a type of inventive problem formulation which is obtained by imposing the following constraints on a given inventive situation: everything remains as is (without any changes) or becomes even simpler, but the required positive effect is provided or the harmful effect disappears.

Comment: Definition of a mini-problem targets at obtaining a solution required with as minimal changes in the existing technical system as possible.

Parameter

Synonym(s): Technical parameter

It is a variable dimensional or dimensionless measurable factor, either specific or aggregated, that participates in the definition of an attribute of a technical system, its subsystem, or supersystem and is expressed in terms related to technology (physical, chemical, etc.).

Comment: Parameters can be physical and non-physical and measured by absolute or relative values.

Examples: Specific level of conductivity, specific level of viscosity measured in Pascal seconds.

Parameter, Typical

Synonym(s): Typical engineering parameter

It is an abstract parameter which is often used to either describe functions of many technical systems (e.g., speed, energy consumption) or used to aggregate a number of more specific parameters (e.g., "complexity," "manufacturability").

Separation of Contradicting Demands

Synonym(s): Separation of conflicting demands

It is a method of solving an inventive problem by satisfying two contradicting demands in the win-win way through separation of the conflicting demands rather than by using optimization or trade-off.

Separation Principles

These are principles of contradiction elimination based on the idea to decouple contradicting demands by their separation.

Comment: Currently, four separation principles are available: (1) separation of contradicting demands in space, (2) separation of contradicting demands in time, (3) separation of contradicting demands in part and a whole, and (4) separation of contradicting demands in condition.

Standard Inventive Problem

It is an inventive problem whose model matches one of the predefined inventive problem models described in the TRIZ tools (primarily, in the system of 76 inventive standards) as well as in the TRIZ knowledge base and which can be solved directly with their use.

Substance

It is a type of discrete matter which possesses rest mass. Substance can be in gaseous, liquid, solid, and plasma states.

Comment: In certain cases and TRIZ tools, the term "substance" can be considered in a broader sense, for example, a material component or an assembly of material components can be called "substance" in the inventive standards or Algorithm of Inventive Problem Solving.

Comment: In substance-field analysis and inventive standards, the use of the term "substance" is not limited to describing types of materials only. It can be a part of a technical system of the entire technical system.

Examples: (1) Water, (2) sand, (3) oxygen, (4) concrete, (5) a cutting tool, and (6) a brick.

Su-Field

Synonym(s): (1) S-field; (2) Su-Fi; (3) Vepol.

It is a model of a minimal technical system which consists of at least two components made of substance and a field providing interaction between the substance components.

Comment: A minimal complete Su-field is presented graphically as a triangle with symbolic nodes depicting the two substances, the field, and lines between the nodes depicting interactions between the components.

Comment: Any technical system can be considered either as a single or as a network of Su-fields. A special type of Su-fields is known as "measurement Su-field" which might include only one substance component, some property of which has to be measured or detected.

Example: A Su-field representing a technical system for hitting a nail with a hammer will include two substances: the hammer, the nail, and one field between them (a "mechanical" field transferring the energy as a result of impact).

Su-Field Analysis

Synonym(s): (1) Substance-field analysis; (2) S-field analysis; (3) Su-Fi analysis; (4) Vepol analysis.

It is a method of abstract symbolic modeling of a technical system or a part of the technical system (sometimes in combination with a part of its supersystem) in terms of substance components, fields, and physical interactions between them.

Comment: In Substance-field modeling, the use of the term "substance" is not limited to describing types of materials only. It can be a part of technical system of the entire technical system.

Su-Field Component

Synonym(s): Su-field element

It is any substance or a field presented as a part of a particular Su-field.

Often, the term "substance component" is replaced with a single word "substance."

Su-Field Diagram

It is a graphic presentation of Su-field in shape of triangle with nodes representing two substances and a field and lines between the nodes representing interactions between the Su-field components.

Comment: The number of nodes and lines may vary.

Transition to Macro-level

It is a method of solving an inventive problem or directly evolving a technical system to the next stage of evolution by either (1) increasing a number of components or interactions forming the system, (2) using several identical or similar system instead of a single system, or (3) assigning dual properties to the system.

Example: A bimetallic plate consists of two metal plates with different coefficients of thermal expansion.

Transition to Micro-level
A method of solving an inventive problem or evolving a technical system by replacing an operational principle behind its subsystem with a new principle that utilizes properties of microscale material objects: fragmented substances, molecules, particles, or physical fields.

Example: Using a laser pointer instead of a wooden stick.

4 TRIZ Tools

Algorithm of Inventive Problem Solving
Abbreviation: ARIZ

Synonym(s): Algorithm of Solving Inventive Problems

It is the central analytical tool of TRIZ (ARIZ is a Russian abbreviation) which consists of a sequence of logical procedures to analyze a vague or ill-defined initial problem/situation and transform it into a distinct system conflict which can be further resolved either independently or with other TRIZ tools.

Comment: Consideration of the system conflict leads to the formulation of a physical contradiction whose elimination is provided with the help of the separation principles and by the maximal utilization of the resources of the subject system. ARIZ is a system of the most fundamental concepts and methods of TRIZ, such as ideal technical system (ideal system), system conflict, physical contradiction, the Su-field analysis, the inventive standards, and the laws of technical systems evolution. The technique includes a number of psychological and systemic operators to support its procedures.

Comment: ARIZ is the acronym abbreviated from Russian term "Алгоритм Решения Изобретательских Задач (Algorithm Reshenya Izobretatelskyh Zadach)" written in Latin letters. In early translations from Russia, ARIZ was abbreviated as AIPS or ASIP.

Contradiction Matrix
It is a matrix which provides a systematic access to the most frequently used inventive principles to resolve a specific type of a technical contradiction. In the contradiction matrix, a specific type of a technical contradiction is selected by the predefined typical engineering parameters.

Comment: The original matrix was developed by G. Altshuller and later updated by other TRIZ developers. Later revisions and modification of the original matrix are usually called "contradiction matrix."

Function Analysis
Synonym(s): (1) Functional analysis; (2) Function-cost analysis; (3) Function-attribute analysis; (4) Value-engineering analysis.

It is an analytical method and a tool to model technical systems and their supersystems in terms of functional carriers, objects of the functions, their functions, and the costs of functions delivery and system components.

Comment: A resulting function model of a technical system helps to better understand, extract, visualize, and categorize functional relationships in the technical system, to rank functions, and identify both existing and potential problems related to the functioning of the technical system.

Inventive Principle

It is a recommendation that provides a generic guideline(s) which indicates how to solve an inventive problem represented as an engineering or physical contradiction.

Comment: Inventive principles were extracted and formulated on the basis of extensive studies of diverse documents describing inventions (such as patents) and innovations. Currently TRIZ knowledge base includes a collection of 40 inventive principles.

Inventive Principle at Macro-level

It is a method of applying an inventive principle without utilizing advantages of scientific effects which can be used to provide a result required by using properties of substances and fields at microscale.

Inventive Principle at Micro-level

It is a method of applying an inventive principle with utilizing advantages of scientific effects which can be used to provide an action required by using properties of substances and fields at microscale.

Inventive Standard

Synonym(s): (1) Standard inventive solution; (2) Standard solution.

It is a problem solving method which provides instruction how to transform a Su-field given which involves a problem to a new Su-field which provides the solution required.

Comment: The description of the instruction consists of two parts: its left part presents an existing Su-field diagram that has to be improved (a generic model of an inventive problem), and its right part presents a Su-field diagram that implements such an improvement (a generic model of a solution).

Comment: A number of inventive standards do not contain Su-field diagrams and use textual explanations instead.

Inventive Problem Analysis

It is a process of transformation of an inventive problem definition given to a well-defined problem formulated in terms of a specific problem model valid for a matching TRIZ tool and extracting information needed to find a solution to the problem.

Comment: In current TRIZ, several alternative methods of inventive problem analysis are available.

Resource Analysis
It is an examination of resources available in a technical system and its supersystem (primarily substances and fields) in order to compile a list of resources that can be used for solving a particular inventive problem.

System of Inventive Standards
Synonym(s): A system of standard inventive solutions

A collection of inventive standards which includes five classes of inventive standards categorized in accord with the particular types of inventive problems the inventive standards can deal with: (1) Completion and decomposition of Su-fields, (2) evolution of Su-fields, (3) transition to macro- and micro-levels, (4) measurement and detection Su-fields, and (5) application of inventive standards.

Comment: Currently, the system of inventive standards contains 76 inventive standards. There are proposals to extend the existing system of inventive standards with newly discovered inventive standards.

Trimming
Synonym(s): Functional idealization

It is a method and a technique for improvement of a technical system by removing (trimming) certain components and redistributing their useful functions among the remaining system or supersystem components while preserving quality and performance of the system.

TRIZ Catalogues of Effects
Synonym(s): TRIZ database of effects

It is a database of scientific effects from a specific scientific discipline in which the effects are structured and categorized according to generic technical functions that can be obtained on the basis of specific scientific effects and phenomena.

Comment: In each catalogue, the effects are combined to different groups which include those effects that can deliver a generic technical function.

Comment: Currently the following catalogues of scientific effects are known: (1) catalogue of physical effects, (2) catalogue of chemical effects, (3) catalogue of geometric effects, and (4) catalogue of biological effects.

5 Laws and Trends of Technical Systems Evolution

Bell Curve of Evolution
It is a curve shaped as a bell depicting a nonlinear relation between costs of resources required to provide the main parameter of value or delivery of a main useful function of a technical system and time during evolution of the technical system.

Comment: The bell curve of evolution includes two phases of a technical system evolution: the phase of expansion (growth of mass, dimensions, and energy consumption required to evolve system's functionality, quality, and performance) which follows by the phase of convolution (reduction of mass, dimensions, and energy consumption required to evolve system's functionality, quality, and performance).

Collection of TRIZ Trends of Technical Systems Evolution

Synonym(s): Collection of TRIZ trends of engineering systems evolution

It is a collection of classical TRIZ which presents 11 trends of technical systems evolution.

Comment: Nine of these trends of technical systems evolution were originally categorized to three groups, (1) trends of statics, (2) trends of cinematic, and (3) trends of dynamics, while the remaining two trends are considered independent.

Comment: In addition to classical TRIZ trends of evolution defined by G. Altshuller, contemporary developers of TRIZ offer a number of additional TRIZ trends of engineering systems evolution.

Directed Evolution

It is a method and a tool for performing the forecast of further evolution of a technical system or a technology with the use of the models, trends, and lines of technical systems evolution.

Law of Technical Systems Evolution

It is the term originated by the founder of TRIZ G. Altshuller to present a number of common generic patterns which were discovered during studying evolution of the vast majority of technical systems, irrespectively, specific engineering domains.

Comment: Currently, for some laws originally defined by Altshuller, the term "law or technical systems evolution" is replaced with the term "trend of technical systems evolution" due to the lack of statistical proof that the laws of technical systems evolution are valid for all technical systems under certain circumstances without exception.

Lines of Technical Systems Evolution

It is a specific line within a trend of technical systems evolution presenting a general direction of evolution which shows a number of specific noncontradictory successive transformations a technical system or its part passes through during its evolution according to a specific criterion.

S Curve of Evolution

It is a curve which represents a nonlinear relation between a change of a value of one of the most important attributes of a technical system and time. The value of the attribute is presented along the vertical axis, while time is presented along the horizontal axis.

Comment: A typical S curve has shape similar to Latin letter "S" due to three distinct time intervals: (1) phase of a system birth and early development where the value of the attribute rises relatively slowly, (2) phase of rapid system growth where the value of the attribute rises rapidly, and (3) phase of maturity where the value of the attribute rises relatively slowly or does not rise at all. Sometimes an extra interval is added to S curve presenting the phase of stagnation.

Comment: S curve is similar to a logistic curve known in other disciplines beyond TRIZ.

Theory of Technical Systems Evolution
Synonym(s): Theory of engineering systems evolution

Abbreviation: TTSE (synonym: TESE).

It is a theory which emerged during evolution of TRIZ and which studies models of evolution of technical systems as well as laws, trends, and lines of the technical systems evolution.

Comment: Sometimes, the early abbreviation is still used: TRTS (from Russian title «Теория Развития Технических Систем» written in Latin characters).

Trend of Technical Systems Evolution
Synonym(s): Trend of technology evolution

It is a line of generic patterns representing a common direction of technical systems evolution through a range of successive transitions of a technical system from one state to another according to a certain domain-independent criteria.

Comment: Some trends of technical systems evolution include more specific lines of technical systems evolution defined according to more narrow criteria.

TRIZ-Based Evolution Forecast
Synonym(s): (1) TRIZ-based technology prediction; (2) TRIZ-based innovation roadmapping.

It is a process of short- or a long-term forecasting and roadmapping further evolution of a technical system given based on using the analytical and problem solving TRIZ tools including the trends and lines of technology evolution in combination with the TRIZ knowledge base.

Comment: In addition to the TRIZ tools and TRIZ knowledge base, during the process of TRIZ-based evolution forecast, external knowledge sources are extensively used from different knowledge areas.

References

Altshuller, G. S. (1984). *Creativity as an exact science*. New York: Gordon & Breach.

Altshuller, G. S. (1993). *To find the idea* (1st ed.). Novosibirsk: Nauka; (2nd ed.): Petrozavodsk: Scandinavia, 2003 [in Russian: Альтшуллер Г.С. Найти Идею, Новосибирск: Наука, 1-е издание, 1986; 2-е издание, 1991; 3-е изд., доп., Петрозаводск: Скандинавия, 2003].

Altshuller, G. S. (1998). *The innovation algorithm: TRIZ, systematic innovation, and technical creativity*. Worcester, MA: Technical Innovation Center.

Altshuller, G. S., Zlotin, B. L, & Zusman, A. V. (1989). *Search for new ideas: From insight towards technology: Theory and practice of inventive problem solving*. Kishinev: Karta Moldavenjaske [In Russian: Альтшуллер Г.С., Злотин Б.Л., Зусман А.В. Поиск новых идей: от озарения к технологии (теория и практика решения изобретательских задач). Кишинев: Картя Молдовеняскэ, 1989].

Altshullers's Foundation: Main terms of TRIZ, version 1.0 [In Russian: Фонд Альтшуллера: Основные термины ТРИЗ/версия 1.0]. Retrieved December 01, 2013, from http://www.altshuller.ru/thesaur/thesaur.asp

Gorin, Y. (1973). *A guide to physical effects for solving inventive problems*. Baku [In Russian: Ю.Горин. Указатель Физических Эффектов для Решения Изобретательских Задач, Баку, 1973].

Khomenko, N. (2013). *TRIZ as a general theory of powerful thinking (OTSM), collection of articles*. 1992–2003. Retrieved December 01, 2013, from http://www.trizminsk.org/e/prs/kho.htm

Korolyov, V. A. (2013). *Glossary of TRIZ-OTSM terms* [In Russian: Королёв В.А. Справочник терминов ТРИЗ-ОТСМ. 13 редакция]. Retrieved December 01, 2013, from http://www.triz.org.ua/enc/

Litvin, S., & Lyobomirskiy, A. (2013). *Laws of engineering systems evolution* [In Russian: Законы развития технических систем. А.Любомирский, С.Литвин]. Retrieved December 01, 2013, from http://gen3.ru/ru/3605/5454/

Litvin, S., Lyrbomirskiy, A., & Sigalovskaya, I. (2013). *G3:ID method glossary* [In Russian: С.Литвин, А.Любомирский, И.Сигаловская. Глоссарий к методике G3:ID]. Retrieved December 01, 2013, from http://gen3.ru/ru/3605/5453/

Litvin, S., Petrov, V., & Rubin, M. (2007). *TRIZ body of knowledge*. TRIZ Summit 2007, St. Petersburg. http://www.triz-summit.ru/en/section.php?docId=3841

Matvienko, N. N. (2013). *TRIZ terms (Problem collection)* [In Russian: Матвиенко Н.Н. Термины ТРИЗ (проблемный сборник]. Retrieved December 01, 2013, from http://www.metodolog.ru/node/689

Official G.S. Altshuller foundation. Retrieved December 01, 2013, from http://www.altshuller.ru

Petrov, V. (2002). *Dictionary of TRIZ terms*, Tel-Aviv.

Petrov, V. (2003). *History of developing standards*. Tel-Aviv [In Russian: В. Петров. История Развития Стандартов. Тель-Авив, 2003].

Petrov, V. (2006). *History of developing the algorithm of solving inventive problems: ARIZ*. Tel-Aviv [in Russan: В. Петров. История разработки Алгоритма Решения Изобретательских Задач. Тель Авив, 2006].

Petrov, V. (2013). *The laws of system evolution*. Berlin: TriS Europe GmbH, 646 pages [In - Russian: INNOVATOR (06) 01/2013, ISSN 1866-4180. Петров В. Законы развития систем. Монография. Тель-Авив, 2013–646 с].

Salamatov, Y. (1999). *TRIZ: The right solution at the right time*. The Netherlands: Insytec.

Souchkov, V. (2014). *The glossary of TRIZ and TRIZ related terms, version 1.0*. The International TRIZ Association (MATRIZ).

Souchkov, V., Kucharavy, D., & Devoino, I. (1991). *Invention machine terms dictionary*, Invention Machine Labs.

Zlotin, B., & Zusman, A. (2001). *Directed evolution: philosophy, theory and practice*. Southfield, MI: Ideation International.

Zlotin, B., Zusman, A., Altshuller, G., & Philatov, V. (1999). *Tools of classical TRIZ*. Southfield, MI: Ideation International.

MIX
Papier aus verantwortungsvollen Quellen
Paper from responsible sources
FSC® C105338

FSC
www.fsc.org

If you have any concerns about our products,
you can contact us on
ProductSafety@springernature.com

In case Publisher is established outside the EU,
the EU authorized representative is:
Springer Nature Customer Service Center GmbH
Europaplatz 3, 69115 Heidelberg, Germany

Printed by Libri Plureos GmbH
in Hamburg, Germany